Edited by
Markus Schmidt

Synthetic Biology

Related Titles

Luisi, P. L., Chiarabelli, C. (eds.)

Chemical Synthetic Biology

2011

ISBN: 978-0-470-71397-6

Parry, S., Dupré, J. (eds.)

Nature After The Genome

2010

ISBN: 978-1-4443-3396-1

Vertes, A., Qureshi, N., Yukawa, H.,
Blaschek, H. (eds.)

Biomass to Biofuels

Strategies for Global Industries

2010

ISBN: 978-0-470-51312-5

Fu, P., Panke, S. (eds.)

Systems Biology and Synthetic Biology

2009

ISBN: 978-0-471-76778-7

Soetaert, W., Vandamme, E. (eds.)

Biofuels

2009

ISBN: 978-0-470-02674-8

Edited by Markus Schmidt

Synthetic Biology

Industrial and Environmental Applications

WILEY-BLACKWELL

The Editor

Dr. Markus Schmidt
Organisation for International
Dialogue and Conflict
Management
Kaiserstr. 50/6
1170 Vienna
Austria

Cover
Birgit Marie Schmidt MA (RCA)
London, UK

Library of Congress Card No.: applied for

British Library Cataloguing-in-Publication Data
A catalogue record for this book is available from the British Library.

Bibliographic information published by the Deutsche Nationalbibliothek
The Deutsche Nationalbibliothek lists this publication in the Deutsche Nationalbibliografie; detailed bibliographic data are available on the Internet at <http://dnb.d-nb.de>.

Wiley-Blackwell is an imprint of John Wiley & Sons, formed by the merger of Wiley's global Scientific, Technical, and Medical business with Blackwell Publishing.

Typesetting Toppan Best-set Premedia Limited, Hong Kong
Printing and Binding Markono Print Media Pte Ltd, Singapore
Cover Design Adam-Design, Weinheim, Germany

Print ISBN: 978-3-527-33183-3
ePDF ISBN: 978-3-527-65926-5
ePub ISBN: 978-3-527-65927-2
mobi ISBN: 978-3-527-65928-9
oBook ISBN: 978-3-527-65929-6

Contents

List of Contributors

Rachel Armstrong
Senior Lecturer, Research & Enterprise
School of Architecture & Construction
University of Greenwich
Avery Hill Campus, Mansion Site,
Bexley Road, Eltham
London SE9 2PQ
UK

Shlomiya Bar-Yam
Massachusetts Institute of Technology
(MIT)
77 Massachusetts Ave., Building E40-437
Cambridge, MA 02139-4307
USA

Mark Bedau
Reed College
3203 SE Woodstock Blvd.
Portland, OR 97202
USA

Jennifer Byers-Corbin
Gryphon Scientific
6930 Carroll Ave, Suite 810
Takoma Park, MD 20912
USA

Rocco Casagrande
Gryphon Scientific
6930 Carroll Ave, Suite 810
Takoma Park, MD 20912
USA

Antoine Danchin
AMAbiotics SAS
Building G1, 2 rue Gaston Crémieux
91000 Evry
France

Florentine Eichler
Massachusetts Institute of Technology
(MIT)
77 Massachusetts Ave., Building
E40-437 Cambridge, MA 02139-4307
USA

Zheng-Jun Guan
Chinese Academy of Sciences
Institute of Botany
20 Nanxincun, Xiangshan
100093 Beijing
China

Allen Lin
Massachusetts Institute of Technology
(MIT)
77 Massachusetts Ave., Building
E40-437 Cambridge, MA 02139-4307
USA

Ismail Mahmutoglu
BAUER Umwelt GmbH
In der Scherau 1
86529 Schrobenhausen
Germany

Martin Österreicher
Massachusetts Institute of Technology (MIT)
77 Massachusetts Ave., Building
E40-437 Cambridge, MA 02139-4307
USA

Kenneth A. Oye
Massachusetts Institute of Technology (MIT)
77 Massachusetts Ave., Building
E40-437 Cambridge, MA 02139-4307
USA

Lei Pei
Organisation for International
Dialogue and Conflict Management
Kaiserstr. 50/6
1070 Vienna
Austria

Manuel Porcar
Universitat de València
Biotechnology and Synthetic Biology
Institut Cavanilles de Biodiversitat i
Biologia Evolutiva
46071 Valencia
Spain

Pernilla C. Regardh
Massachusetts Institute of Technology (MIT)
77 Massachusetts Ave., Building E40-437
Cambridge, MA 02139-4307
USA

Vincent Schachter
Total Gas & Power
Research and Development
2, place Jean Miller – La Défense 6
92078 Paris La Défense Cedex
France

Markus Schmidt
Organisation for International
Dialogue and Conflict Management
Kaiserstr. 50/6
1070 Vienna
Austria
and
Biofaction KG
Grundsteingasse 36/41
1160 Vienna
Austria

Helge Torgersen
Institute of Technology Assessment
Austrian Academy of Sciences
Strohgasse 45, 5
1030 Vienna
Austria

Ralph D. Turlington
Massachusetts Institute of Technology (MIT)
77 Massachusetts Ave., Building E40-437
Cambridge, MA 02139-4307
USA

Wei Wei
Chinese Academy of Sciences
Institute of Botany
20 Nanxincun, Xiangshan
100093 Beijing
China

Short CV of Contributors

Rachel Armstrong is a medical doctor with qualifications in general practice, a multi-media producer and an arts collaborator whose current research explores the possibilities of architectural design to create positive practices and mythologies about new technology. She is collaborating with international scientists and architects to explore cutting-edge, sustainable technologies by developing metabolic materials in an experimental setting.

Shlomiya Bar-Yam is a graduate student in the Technology and Policy Program at MIT. She is working on the environmental implications of synthetic biology in release contexts. She graduated from Brown University with a BSc in Biology with a focus in ecology, and she has worked on sustainability programs and at the New England Complex Systems Institute as a science writer.

Mark A. Bedau is an internationally recognized leader in the interdisciplinary study of complex adaptive systems. He co-founded and is currently COO of ProtoLife Srl in Venice, Italy. He also cofounded the European Centre for Living Technology (UNIVE), in Venice, Italy. He is Professor of Philosophy and Humanities at Reed College and visiting Professor at the European School of Molecular Medicine (Milan, Italy). He is internationally recognized as a uniquely qualified expert in the philosophical foundations of complex adaptive systems.

Jennifer Byers-Corbin is a health effects and medical countermeasure modeler at Gryphon Scientific. Dr. Corbin received her PhD in biomedical science from New York University. Since then, she has applied her training to the study and analysis of biosafety and security. She represents BARDA modeling on several working groups on a broad range of topics related to biodefense. Dr. Corbin is an able and effective leader who has served as principal investigator or task leader for multiple projects for United States government clients.

Rocco Casagrande (BA chemistry and biology, Cornell University; PhD experimental biology, MIT) is the Managing Director of Gryphon Scientific, LLC. Over the past several years, Dr. Casagrande led several projects related to weapons of mass destruction and science policy for various United States Federal and State agencies. These projects include WMD threat and risk assessments and biodefense system evaluations, technology assessment, modeling and technical guidance. From December 2002 to March 2003, Dr. Casagrande served as an UNMOVIC biological weapons inspector in Iraq where he obtained hands-on experience with chemical and biological agents.

Antoine Danchin is a French geneticist known for his research in several fields of biology. Originally he was trained as a mathematician at the Institut Henri Poincaré and as a physicist at the Ecole Polytechnique. He is the Chairman of the startup AMAbiotics, specialized in metabolic bioremediation and synthetic biology. He was the director of the Department Genomes and Genetics at the Institut Pasteur in Paris where he headed the Genetics of Bacterial Genomes Unit.

Florentine Eichler has a graduate degree in law from the University of Vienna, Austria. During her studies in Austria and the Netherlands she focused on medical law. She is currently a Research Associate at the MIT Program on Emerging Technologies (PoET).

Zheng-Jun Guan received her PhD on cell biology from the College of Life Sciences, Northwest University, China. She was a teacher of cell biology for eight years. She is now working as a postdoctoral scientist on the biosafety of synthetic biology and genetically modified organisms at the Institute of Botany, Chinese Academy of Sciences.

Allen Lin received a BS in Chemical–Biological Engineering and a BS/MEng in Electrical Engineering and Computer Science in 2011 from MIT, and he is currently a research technical assistant in synthetic biology at the Weiss Laboratory at MIT. Since 2009, he has worked with Prof. Kenneth Oye on synthetic biology risk assessment and management.

Ismail Mahmutoglu is a Chemist at Bauer Umwelt GmbH, a specialist on the remediation of brownfields and the treatment of waters and gases. He is a specialist for the design and manufacture of water treatment plants for decontamination, for waste water and for potable water. The range of projects he is dealing with includes complex technical treatment steps with biological treatment steps, but also *in situ* technologies to improve the underground conditions for the microorganism.

Kenneth A. Oye is Director of the MIT Program on Emerging Technologies with a joint appointment in Political Science and Engineering Systems. He is an NSF SynBERC PI and an iGEM judge and biosafety coordinator. He serves on the NRC Board on Global Science and Technology and the Advisory Committee for the International Risk Governance Council. Recent publications include "Adaptive licensing," (*Nature CPT*, forthcoming), "Synthetic biology and the future of biosecurity" (*Politics and Life Sciences*, 2010), "Planned adaptation in risk regulation"

(*Technology Forecasting and Social Change*, 2010), "Intellectual commons and property in synthetic biology" (*Synthetic Biology*, 2009) and "Embracing uncertainty" (*Issues in Science and Technology*, 2009).

Martin Österreicher has a graduate degree in law from the University of Vienna. He is currently working as a research associate at the MIT Program on Emerging Technologies (PoET). In the course of his studies in Vienna and St. Gallen, Switzerland, he focused his research on international law and genetic law.

Lei Pei completed her PhD at the Division of Clinical Bacteriology, Karolinska Institute, Sweden, in 2002. After her PhD she worked at the Division of Infectious Diseases, Department of Medicine, at the Massachusetts General Hospital/ Harvard Medical School, Boston, Massachusetts, as a Postdoctoral Research Fellow. Between 2005 and 2009 she completed her second postdoc position at the Flanders Institute for Biotechnology, Department of Molecular Biology, Gent University, Belgium. Since 2009 she has been working with Markus Schmidt as a postdoc on synthetic biology and risk assessment.

Manuel Porcar is an applied microbiologist and biotechnologist. He leads the Biotechnology and Synthetic Biology Laboratory at the Cavanilles Institute of the University of Valencia, where he coordinates several research efforts focusing on strain and gene selection strategies for bioremediation, biofuels production and bioenergetics. As an expert on GMOs evaluation, he is a member of the Comisión Nacional de Bioseguridad, the Spanish reference organ on GMOs. He has been actively involved in the international Genetically Engineered Machine (iGEM) competition and in the development of devices for synthetic biology. He is a convinced supporter of Darwinian approaches as a tool for synthetic biology.

Pernilla Regardh holds an MSc in Technology and Policy from the Massachusetts Institute of Technology, a BSc in Biotechnology from the Royal Institute of Technology in Stockholm, Sweden, and a BA in Political Science from Stockholm University. Her research interests include risk assessment, regulatory development and public debate over emerging biotechnologies, and as a graduate student at MIT she analyzed and compared the regulatory landscape of synthetic biology in Europe and the United States together with Dr. Kenneth Oye. Regardh recently graduated from MIT and is currently working as a strategy consultant in Stockholm.

Vincent Schachter is Vice-President for Research and Development at Total Gas and Power. Before that he was the director of Systems Biology of the French CEA, where he lead the Computational Systems Biology research group. He holds a PhD in Computer Science from the Ecole Normale Supérieure in Paris and entered bioinformatics through the field of protein interaction network analysis. He has also acquired applied experience with high-throughput experimental data – protein–protein interactions, sequence, cellular phenotypes – first as

Director of Bioinformatics Research at Hybrigenics SA, a biotech company, and then as Director of Bioinformatics at CEA. He is a cofounder of the BioPathways Consortium and a participant in the BioPAX standardization effort and is a referee for several bioinformatics and biology journals.

Markus Schmidt has an interdisciplinary background with an education in electronic and biomedical engineering (Ing), biology (MSc) and environmental risk research (PhD). His research interests include the risk assessment, science–society interface and technology assessment (TA) of novel bio-, nano- and converging technologies. Since 2005 he pioneered synthetic biology safety and ethics research in Europe. He is co-founder and board member of the Organization for International Dialog and Conflict Management (IDC) and is co-founder and CEO of Biofaction. For details, see www.markusschmidt.eu.

Helge Torgersen was a researcher and lecturer at the Institute of Molecular Biology and the Institute of Biochemistry, University of Vienna from 1981 to 1989. In 1990 he joined the Institute of Technology Assessment (ITA) of the Austrian Academy of Sciences as a researcher. His main interests are comparative biotechnology policy and safety regulation, the risk assessment and public perception of transgenic organisms, science studies in biotechnology and methods of participatory technology assessment. Current interests include societal aspects of nanotechnology, genomics and synthetic biology.

Ralph D. Turlington is a graduate student in the Technology and Policy Program at MIT working with Dr. Kenneth Oye on the security implications of synthetic biology. He became interested in security during his first two years of college, spent at the United States Military Academy at West Point. Ralph graduated with a BSc in Environmental Science with a second major in Economics from the University of Virginia. His undergraduate research was on atmospheric science and pollution transport studies in the Nepal Himalayas.

Wei Wei received his PhD on Botany from the Institute of Botany, Chinese Academy of Sciences. His major research interests are in plant ecology and biodiversity conservation. He has worked on the biosafety issues of genetically modified organisms for a decade. He is now interested in studying the biosafety link between genetic engineering and synthetic biology (SB) and aims to develop proper risk assessment and management strategies of SB. He is member of the Ad Hoc Expert Group on Risk Assessment and Risk Management within the framework of the Cartagena Biosafety Protocol.

Preface

During the first decade of the 21st century the establishment of Synthetic Biology (SB), a science and engineering field that wants to turn biology into a true technology, could be observed. SB goes beyond previous efforts that use more conventional or, as some consider them, "artisan" forms of biotechnology. With the aim of making biology accessible to the needs of everyday life, SB applies engineering principles such as standardization, modularization, using hierarchies of abstraction or the decoupling of design and fabrication to biological systems in order to establish a whole new set of applications for society.

The aim of this book is to give a critical 360-degree assessment of a selected number of highly promising industrial and environmental applications enabled by SB. The assessment not only analyses to what extent SB could improve current technologies, it also approaches the potential applications from the economic, environmental, social and ethical perspective. The book summarizes these different viewpoints in order to present a balanced evaluation of the technical, economic, environmental and societal ramifications of SB applications.

Chapter 1 provides a detailed analysis of several different types of biofuel and their production, such as ethanol, non-ethanol-based fuels, fuels made from algae, biohydrogen, microbial fuel cells and bio-photovoltaic systems. It describes the technical limitations of ethanol usage and the conditions under which its production would be environmentally sustainable. It also discusses the benefits and production challenges of butanol, further the specific economic context necessary for algae production systems in order to be competitive, and the uncertainty of the environmental impact of open pond production systems. It also takes a closer look at the environmental benefits and infrastructure problems of the biohydrogen economy and finally describes the niche markets for environmentally friendly microbial fuel cells.

Chapter 2 details a number of applications for bioremediation. It focuses on possible methods of how to detect environmental pollutants, views the environmental and social benefits of water treatment, soil and groundwater decontamination, and describes the market opportunities in solid waste treatment. Furthermore, Chapter 2 reports on the challenges of one of the most important applications of SB in the light of future global water scarcity, namely, water desalination by using special biomembranes. Another global environmental challenge that might be turned into a business opportunity is SB-enabled CO_2 recapture.

Chapter 3 is devoted to the assessment of SB-produced biomaterials, such as biopolymers (bioplastics), bulk chemicals or fine chemicals. This chapter highlights the environmental benefits of SB that are reached by leaving behind synthetic chemistry which often comprises an energy-intense production system including toxic side products that could altogether be avoided by using cell factories designed by SB. Another focus in this chapter is devoted to cellulosomes that can be used to degrade cellulose and hemicellulose, and the tremendous impact its use may have on the economy and the environment.

Chapter 4 explores two promising SB technologies that are still at an experimental stage. Both protocells and xenobiological systems could one day provide a whole new set of applications, ranging from smart semi-living systems in the environment to the installation of a genetic firewall in order to impede horizontal gene-flow between natural and engineered organisms, touching on a number of environmental and ethical questions.

The final chapter, Chapter 5, takes the reader on a guided tour through the regulatory frameworks applied to SB. Acknowledging regional differences, this chapter provides insight into the situation in the USA, Europe and China. In addition to laying out the existing regulatory framework, Chapter 5 also suggests improvements and adaptations in order to deal with upcoming technical inventions.

My special thanks go to all the contributors to this book, who provided their expertise coming from academia, industry and non-profit organisations. I would like to thank Wiley for kindly accepting the publication of this book. Last but not least, I would also like to thank you, the reader, for your interest in the scientific, economic, environmental, social and ethical consequences of upcoming SB applications.

I hope that this book contributes to a broader understanding of the many societal ramifications of SB and helps in making smarter decisions about upcoming SB tools and applications.

Markus Schmidt

Acknowledgments

Markus Schmidt, Ismail Mahmutoglu, Manuel Porcar, Rachel Armstrong, Mark Bedau and Lei Pei gratefully acknowledge the financial support provided by the European Commission's seventh framework programme: "TARPOL Targeting environmental pollution with engineered microbial systems à la carte", 2008–2010, FP7 EU-KBBE-212894. Antoine Danchin gratefully acknowledges the support provided by the Fondation Fourmentin-Guilbert and the EU's 7th framework programme "Microme" grant KBBE_2007_3-2_08_222886_2. Markus Schmidt, Lei Pei, Wei Wei and Zheng-jun Guan gratefully acknowledge the financial support provided by the FWF (Austrian Science Fund) and the NSFC (National Natural Science foundation of China) joint-project "Investigating the biosafety and risk assessment needs of synthetic biology in Austria (Europe) and China," project number I215-B17 and NSFC 30811130544. Markus Schmidt gratefully acknowledges the financial support provided by the FWF (Austrian Science Fund) project "SYNMOD: Synthetic biology to obtain novel antibiotics and optimized production systems," project number I490-B12, through the EUROSYNBIO Programme of the European Science Foundation. The NSF Synthetic Biology Engineering Research Center (SynBERC) provided financial support for Shlomiya Bar-Yam, Jennifer Byers-Corbin, Rocco Casagrande, Florentine Eichler, Allen Lin, Martin Österreicher, Pernilla C. Regardh, Ralph D. Turlington, Kenneth A. Oye under NSF Grant 050869. Thanks to Michael Stachowitsch for English proofreading of the Introduction and Chapters 1 to 4.

Executive Summary

Markus Schmidt

This book provides a glimpse into the future of synthetic biology (SB) and its potential applications in the area of environmental and industrial biotechnology. There are a number of applications where SB could well make a difference in making society more economically and environmentally sustainable. This report highlights four major areas (biofuels, bioremediation, biomaterials, novel developments in SB) with a total of 20 applications where SB has a great potential to improve currently available technologies. Each of the 20 applications has been assessed in detail in order to determine: (i) to what extent SB could improve current technologies, (ii) what the economic impact of SB could be, (iii) what the environmental benefits and downsides could be and (iv) whether any social or ethical problems would be created, exacerberated or improved. This assessment is intended to support not only researchers and students, but also national and international funding agencies in their decisions to allocate resources to SB-based biotech applications while taking into account any foreseeable economic, environmental and social/ethical issues. Our outlook is based on state of the art science, although there is clearly a considerable degree of uncertainty about future development paths. This uncertainty needs to be acknowledged when providing recommendations for what we see as the most promising directions for SB in environmental biotechnology. Another area where uncertainties might be ahead, is the *regulatory framework* such as laws, regulations, guidelines and code of conducts, for synthetic biology. As of 2011, hardly any specific regulation has been produced for SB, for the moment current regulations–originally put in place to deal with genetic engineering–cover SB more or less sufficiently well. This situation, however, will change sooner or later, requiring future adaptations to the regulatory status quo.

Biofuels

We are convinced that synthetic biology can help to produce state of the art and next generation biofuels. Current efforts are mainly targeted towards an improved production of *bio-ethanol* from agricultural products, although this approach harbors significant problems because ethanol exhibits certain technical drawbacks (miscible with water, limited use in existing engines). Other non-ethanol biofuels such as *bio-butanol* or *biodiesel* are much better suited to replace petroleum-based gasoline, as their chemical properties resemble it much closer. Synthetic biology could help to overcome current impasses in the production of butanol and other non-ethanol fuels, namely poor fermentation yield and toxicitiy to butanol-producing microorganisms. One shortcoming faced by most biofuels produced from plant material is limitations in the use of hemi- and lignocellulosic material. Any improvement in that area would increase the economic feasibility of biofuel production. One important problem will arise should synthetic biology be able to help solve the above technical issues, namely that more and more agricultural land will be devoted to plant energy crops instead of food crops. In order to avoid such competition with food, we suggest also using non-food-competing biological resources such as perennial plants grown on degraded lands abandoned for agricultural use, crop residues, sustainably harvested wood and forest residues, double crops and mixed cropping systems, as well as municipal and industrial wastes.

In addition to agricultural-based ethanol, biodiesel and butanol, *algae-based biofuels* and *biohydrogen* also deserve consideration. Current concepts foresee a significant advantage of algae-based over agriculture-based biofuels because of higher yield per area and an independence from arable land and clean water. Initial calculations, however, predict that future algae production systems will be economically feasible only if the price for one barrel oil is consistently above US$ 100 and if the production systems entail an area of at least 200 ha. The capital costs of such large production facilities will probably lead to an exlusion of small and medium enterprises (SMEs) and favor "big oil" (or "big energy") companies. Still, algae production systems could be a highly promising avenue of future fuel production once major obstacles such as algae genomics, metabolism and harvesting are overcome. Although *bio-hydrogen* has been praised as an extremely promising fuel by many scientists, our assessment is more cautious. Hydrogen is only useful as fuel if large changes in infrastructure take place (distribution and storage systems, new fuel cell engines). This points to a more distant future beyond 2050, also termed as the hydrogen economy. Although synthetic biology could well help improve the yield of hydrogen-producing cyanobacteria, the actual impact of hydrogen in society and economy depends much more on other areas such as infrastructure. Finally, we analyzed the prospects of *microbial fuel cells* (MFC) as an energy converter. Although we see MFCs as extremely promising and an area where synthetic biology could contribute significantly, it will most likely be applied only in certain niche markets and selected areas of application, rather than large-scale deployment due to the limited energy production.

Bioremediation

Bioremediation is an area with a great potential of benefits provided by synthetic biology. Bioremediation is usually applied on materials with a massive occurence such as *solid (organic) wastes, sewage, industrial waste water, contaminated soil or contaminated groundwater*–all measured in millions of tons or cubic meters. We believe that SB has the potential to create tools to improve the treatment methods, saving costs and environmental resources. Moreover, it can provide methods to produce energy or valuable goods from waste or wastewater. It can also provide tools for producing *fresh or drinking water* from either contaminated water or seawater. Another possible field of application is the production of biosensors to monitor environmental goods and hazards. In a differentiated evaluation, we conclude that biosensors provided by SB tools would have a great positive effect on the environment because they can help to survey environmental hazards more precisely and effectively. Their economic and social impact, however, is rather low due to their niche-product status.

Synthetic biology-based approaches may provide a way of *capturing, storing and recycling carbon dioxide*. This may be through the re-engineering of existing organisms or the creation of novel carbon processes, especially using bottom up approaches where inorganic chemistry is linked to living processes through agents, such as the emerging protocell technology.

Synthetic biology-based carbon capture may not be able to sink sufficient carbon dioxide to completely remediate the currently escalating levels being released through fossil fuel consumption. This is because very large scale geoengineering-scale approaches would be necessary. This approach, however, does offer the possibility of carbon capture and recycling, which current industrial-scale processes cannot do.

Our recommendation is that, because of the scale of the problem with carbon dioxide emissions and the urgent need for remediation, synthetic biology approaches be supported in order to develop the next generations of carbon capture technologies. These will do more than merely store the carbon dioxide but will recycle it into fuels and biopolymers with positive environmental impact.

Another positive impact, particularly to the environment, can be expected for *soil and groundwater remediation*, especially with regard to enhancing the clean-up efficieny and the development of new methods. In this field, the economic and social impacts are rather moderate because it is a specific field with a limited scope of time.

We expect the *strongest impact* in *solid waste and wastewater treatment* and for *water desalination*. The importance of the latter cannot be overstated in a world where billions of people have no access to clean drinking water or to freshwater for agricultural use. Solid waste and wastewater treatment also bear a great potential for improvement by synthetic biology due to their sheer amount and their considerable organic content. We therefore strongly recommend supporting the development of these applications. One possible constraint deserves to be mentioned: solid waste and waste water cannot be treated in sealed vessels or rooms, simply due to their huge volumes. They have to be treated openly in piles or basins. Therefore, the use of engineered cells may create interactions with the environment, a problem that must be kept in mind. At the same time, we expect no limitations in the use of non-proliferative systems such as enzymes or protocells created with the aid of synthetic biology.

Biomaterials

Synthetic biology will have a significant impact on the biomaterials market, particularly in the areas of fine chemicals and bioplastics. We recommend a tool box of products that will act as biodegradable materials. The *bulk chemicals industry* will also be significantly affected by synthetic biology-based technology. The acceptance of this approach and therefore its environmental impacts will be slower, although when new practices are adopted, changes will last longer and take place on a much larger scale. In the *fine chemicals industry*, the incentives for investment relate to the economic potential of the end product (in contrast to bulk chemical manufacturing). The payoffs could have environmental benefits, although these may be largely limited to more efficient use of energy because the core manufacturing practice relies on petrochemicals. There is also limited potential for synthetic biology-based techniques to help avoid recalcitrant molecules in the production process. Accordingly, investment in synthetic biology-based processes in bulk chemicals is likely to have a positive overall effect on the manufacturing systems used in running the plants. These include the use of biodiesel and less overall chemical waste. Moreover, the large scale at which these processes take place mean that small changes may have significant positive environmental effects. For both fine and bulk chemical production we recommend deploying the "chemical building block system" as designed by (or similar to) the US Department of Energy (DoE).

The field of biopolymers and *bioplastics* urgently needs revisiting in terms of its current labeling for recycling purposes. This is because categorization of the various products is extremely complex, with negative economic consequences (bioplastics are not necessarily biodegradable). We recommend applying a method that renders recognizable those bioplastics that need recycling and those that can be composted, before introducing large-scale use of synthetic biology for bioplastics production. There is an urgent need to develop completely biodegradable plastics; this would benefit from focused synthetic biology research and development in this field. Additionally there is a pressing need for *high-performance structural bioplastics* for manufacturing coupled with completely biodegradable additives. Both of these significant growth areas in the bioplastics industry could be greatly improved by synthetic biology-based research.

Investment is particularly needed in research and development for new methods and products that will expand and develop tools and manufacturing processes with reduced environmental impact compared with the current manufacturing approaches. Here, adequate biosafety constraints on large-scale manufacturing units need to be established. *Cellulosomes* (complex molecules that degrade hemi- and lignocellulosic material) possess high economic potential for biofuels, paper and waste processing. Synthetic biology has the potential to design more efficient and completely new cellulosome complexes, yielding new, efficient cellulose-digesting proteins. We recommend open sourcing of the cellulosome technology due to the justice of distribution issues involved in the technology.

Novel Developments in Synthetic Biology

Protocell technology represents a bottom up approach to synthetic biology that bridges inorganic and organic processes. Protocell technology promotes a better understanding of synthetic biology as a whole to develop new technologies. Although the research is in an early stage, the development of potentially radically novel and significant environmental interventions is feasible. Examples include remediation of carbon emissions and alternative biofuels technology. One strong recommendation is the investment in basic science to underpin and support the research while private investment is being geared up.

Protocell technology has a huge potential to offer tools and methods radically different from those previously encountered with synthetic biology approaches. This reflects its bottom up nature and its overlap with basic chemistry. This also calls for looking into a toolbox of potential products and for investigating issues related to make the technology open source. *Xenobiology* (also known as chemical synthetic biology) is another bottom up approach to design and construct radically new biological systems with properties not found in nature. Using non-canonical amino acids, alternative base pairs to enlarge the genetic alphabet, or different chemcial backbones in a xenonucleic acid, these chemically modified organsims and systems will enable a much higher level of biosafety when using engineered biosystems for, or in, the environment. For example, novel enzymes (such as amylase) with non-canonical amino acids can be used to reduce the optimal temperature for breaking starch down into glucose. This would save enormous amounts of energy and help reduce greenhouse gas emissions. This would help reduce greenhouse gas emissions. Organisms with an enlarged genetic alphabet or a DNA with a different chemical backbone could be designed by synthetic biology. This approach would impede horizontal gene transfer and genetic pollution between engineered and natural organisms. Similar to protocells, xenobiology is in a very early stage of development and requires increased support for basic research in order to introduce radically new concepts and applications.

Table A summarises our assessment of each of the 20 applications. For further details see the following chapters.

Table A Assessment of potential synthetic biology applications in environmental biotechnology over the next 10–15 years.

	Opportunity for SB contribution	**Economic benefit**	**Environmental impact**	**Ethics and social impact**	**Overall assessment**
Biofuels					
Ethanol	◉	◉	◉	⭘	◉
Non-ethanol	●	◉	◉	⭘	◉
Algae-based fuels	⭘	●	◉	⭘	◉
Bio-hydrogen	⭘	◉	●	◉	⭘
Microbial fuel cell	◉	⭘	⭘	◉	◉
Bioremediation					
Biosensors	●	⭘	●	◉	●
Soil and ground-water remediation	○	◉	●	⭘	⭘
Water desalination	⭘	●	◉	●	●
Soil remediation	⭘	⭘	◉	◉	⭘
Solid waste	◉	◉	◉	⭘	◉
CO_2 recapturing	◉	⭘	◉	◉	◉
Biomaterials					
Bioplastics	●	●	◉	◉	●
Bulk chemicals	◉	◉	◉	◉	◉
Fine chemicals	●	●	●	○	●
Cellulosomes	◉	●	◉	○	◉
Novel developments					
Protocells	●	○	⭘	⭘	◉
Xenobiology	●	⭘	◉	⭘	◉

Key:
○ Low or even negative impact, no or hardly any improvement.
⭘ Rather low impact, some improvement.
◉ Positive impact, notable improvement.
● Excellent, significant improvement.

Introduction

Markus Schmidt

The following report provides a glimpse into the future of synthetic biology (SB or synbio) and its potential applications in the area of environmental biotechnology. There are a number of applications where SB could well make a difference in order to make society more economically and environmentally sustainable. In this report we highlight four major areas (biofuels, bioremediation, biomaterials and novel developments in SB) with a total of 20 applications (3 general areas and 17 specific applications) where SB can probably improve currently available technologies. Each of the 20 applications has been assessed in detail in order to determine to what extent SB could improve current technologies. We asked:

- What could be the economic impact of SB?
- What could be the environmental benefits and downsides?
- Would any social or ethical problems be created, exacerberated or improved?

This assessment is intended not only to guide scientists and students when selecting their next research topic, but also to support national and international funding agencies in their decisions to allocate resources to SB-based biotech applications while taking into account any foreseeable economic, environmental and social/ethical issues.

What Are Synthetic Biology Applications?

Although the term "synthetic biology" was already in use about 100 years ago – in 1910 and 1912 by Stephane Leduc (Campos, 2009) – the contemporary version is a relatively young field at the intersection of biology, engineering, chemistry and information technology. Not atypical for an emerging science and engineering field, a variety of definitions are circulating in the scientific community, and no one definition would receive total support by the researchers involved in SB activities. The probably least contested definition is that found on the SB community webpage (http://syntheticbiology.org/):

Synthetic Biology: Industrial and Environmental Applications, First Edition. Edited by Markus Schmidt.

"Synthetic Biology is:

A) the design and construction of new biological parts, devices, and systems, and

B) the re-design of existing, natural biological systems for useful purposes.

Synthetic biologists are currently working to:

- specify and populate a set of standard parts that have well-defined performance characteristics and can be used (and re-used) to build biological systems,
- develop and incorporate design methods and tools into an integrated engineering environment,
- reverse engineer and re-design pre-existing biological parts and devices in order to expand the set of functions that we can access and program,
- reverse engineer and re-design a "simple" natural bacterium,
- minimize the genome of natural bacteria and build so-called protocells in the lab, to define the minimal requirements of living entities, and
- construct orthogonal biological systems, such as a genetic code with an enlarged alphabet of base pairs."

Activities that fall under SB are currently performed in several (sub)fields. For various reasons, these activities are not always addressed under the term proper, while others use the label for still different endeavors. By and large, however, the following activities are usually subsumed under SB (Bedau *et al.*, 2009; Benner and Sismour, 2005; Deplazes, 2009; Luisi, 2007; O'Malley *et al.*, 2008; Schmidt *et al.*, 2009):

- DNA synthesis (or synthetic genomics);
- Engineering DNA-based biological circuits (based on metabolic engineering but using real engineering principles);
- Defining the minimal genome (or minimal cell);
- Building protocells (or synthetic cells);
- Xenobiology (aka chemical synthetic biology).

Which Synthetic Biology Applications Did We Consider?

The aim of this study was to tackle environmental pollution using synthetic biology. Examining the different fields of biotechnology reveals several color-coded subcategories, defined by their major area of application: **red** biotechnology, that is applied to medical processes; **green** biotechnology, that is biotechnology applied to agricultural processes; **white** biotechnology, also known as industrial biotechnology, is biotechnology applied to industrial processes; **brown** (**gray**) biotechnology is also known as environmental biotechnology and involves different forms of biodegradation and sustainable production; and **blue** biotechnology, that is the marine and aquatic applications of biotechnology.

In addition to these frequently used terms, we can also find **yellow** biotechnology, which is sometimes used as a synonym for the venous (recycling) biotech industry, and **black** biotechnology, which describes the use of biotech for bioweapons or bioterrorism.

For our work, two biotechnology areas were of greater interest: environmental biotechnology (including bioremediation, green manufacturing processes, sustainable development) and industrial biotechnology, which aims to prevent pollution, conserve resources and reduce costs. Bioremediation is uniquely assigned to environmental biotechnology, but what is described as sustainable development or green (environmentally friendly) manufacturing processes can be found in both categories.

Selecting and Assessing Synthetic Biology Applications

The categories of environmental and industrial biotechnology encompass a variety of applications throughout different industrial sectors, as shown in Table 1.

After identifying a range of applications like the ones shown in Table 1, we selected 20 (potential) SB applications that we estimate: (a) have a considerable impact on the economy, environment and/or society and (b) could be available on the market within the next 15 years or lay the foundation for promising new applications. We then selected the 20 applications for further analysis (see Table 2). We then asked the following questions for each of the 20 selected applications:

- How could SB improve or enable the application?
- What is the foreseeable economic impact in case SB improves improves/ enables the application?
- What will be the most likely environmental impact?
- What are potential social and ethical aspects that we need to consider?

In the last step, we prepared recommendations based on our 360-degree analysis (based mainly on literature analysis) of the economic, environmental, societal and ethical impact of selected synbio applications. The report benefits from the fact that the authors come from different backgrounds including non-governmental organizations, universities, small and medium enterprises, and large enterprises. The structure of the report and its recommendations reflect our understanding of the management of new and emerging science and technologies: we sign up neither to an "anything goes" nor a "stop everything" regime, but try to assess different technologies and applications case by case in a more balanced mode of operation (see Figure 1).

Table 1 Examples of environmental biotech industry applications.[a)]

Fine chemical production (e.g., pharmaceutical products, cosmetics) and biohydrogen production	
Bulk chemical production	Chem/bio-warfare agent decontamination
Bio-ethanol transportation fuels	Pulp and paper bleaching
Chiral compound synthesis	Specialty textile treatment
Pharmaceutical manufacturing/processing	Antibiotic production
Synthesis of vanillin and other food flavoring agents	Electroplating/metal cleaning
Biopolymers/plastics	Nutritional oil production
Rayon production	Sweetener production
Metal refining	Oil well drill hole completion
Oil and gas well completion	Textile dewatering
Vitamin production	Vegetable oil degumming
Coal, oil and gas desulfurization	Biological fuel cells
Leather degreasing	Soil decontamination (recalcitrant compounds)
Control of biofilms	Solid waste treatment (industry, mechanical-biological treatment)
Bio-solar panels	Water treatment (groundwater, industrial, domestic, land fill leachate)
More efficient biogas production facilities	Air treatment (biofilter)

a) See: http://www.bio.org/ind/pubs/cleaner2004/ and http://www.bio.org/ind/background/thirdwave.asp.

Although our assessment is based on the scientific state of the art, there is of course a notable degree of uncertainty about future development paths. We have to acknowledge the uncertainties when giving recommendations for what we see as the most promising directions for SB in environmental biotechnology.

The Regulatory Context for Synthetic Biology

In the final three chapters of this book, we take a look at the state of the art regulatory environment in which SB develops. The analysis of the legal situation in the United States, Europe and China shows that almost the entire field of SB is dealt with by a set of regulations, laws, and guidelines originally designed to deal

Table 2 List of selected SB applications.

1 Biofuels
1.1 Biofuels in general
1.2 Ethanol
1.3 Non-ethanol fuels
1.4 Algae-based fuels
1.5 Hydrogen production
1.6 Microbial fuel cell (mfc) and bio-photovoltaic

2 Bioremediation
2.1 Bioremediation in general
2.2 Detection of environmental pollutants (biosensors)
2.3 Water treatment
2.4 Water desalination with bio-membranes
2.5 Soil and groundwater decontamination
2.6 Solid waste treatment
2.7 Carbon dioxide recapturing

3 Biomaterials
3.1 Biomaterials in general
3.2 Biopolymers/plastics
3.3 Bulk chemical production
3.4 Fine chemical production
3.5 Cellulosomes

4 New developments in synthetic biology
4.1 Protocells
4.2 Xenobiology

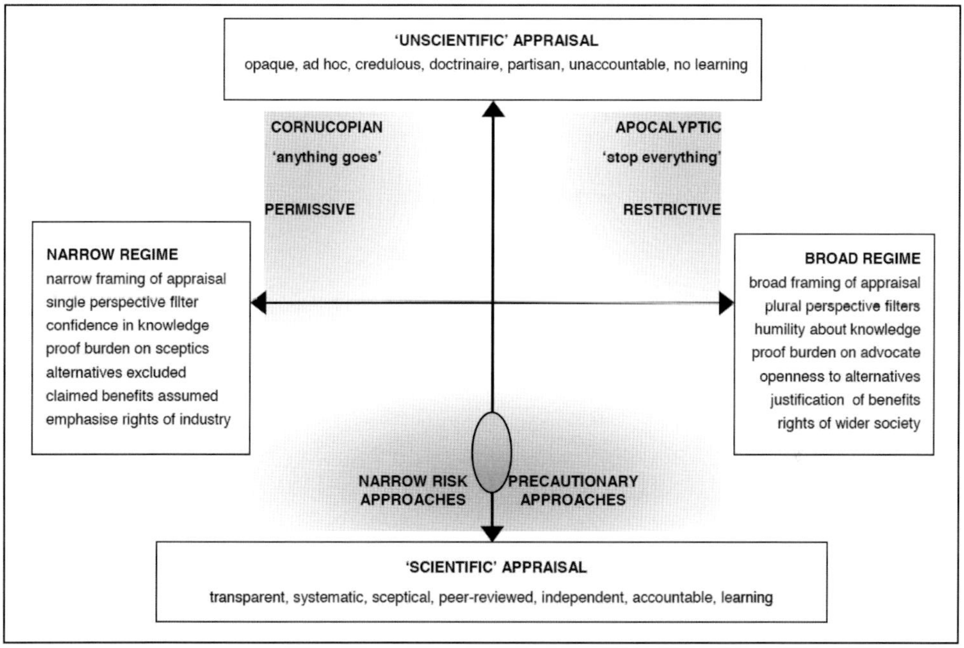

Figure 1 A model of the relationships between the concepts of risk, science and precaution when new technologies are assessed. The elipse represents the intended approach chosen by the authors of this report. (Source: Sterling, 1999).

with "traditional" genetic engineering. Only a few guidelines or self-imposed code of conducts have been prepared specificially for SB, especially regarding biosecurity risks related to DNA synthesis. Although the current set of regulations appear to cover most of what SB is doing (in 2011), it is foreseeable that, in the not so distant future, a number of adaptations need to be made to deal with novel developments.

References

Bedau, M.A., Parke, E.C., Tangen, U., and Hantsche-Tangen, B. (2009) Social and ethical checkpoints for bottom-up synthetic biology, or protocells. *Syst. Synth. Biol.*, **3**, 65–75.

Benner, S.A., and Sismour, A.M. (2005) Synthetic biology. *Nat. Rev. Genet.*, **6**, 533–543.

Campos, L. (2009) That was the synthetic biology that was, in *Synthetic Biology: The Technoscience and Its Societal Consequences* (eds M. Schmidt, A. Kelle, A. Ganguli-Mitra, and H. deVriend), Springer, Chapter 2, pp. 40–48.

Deplazes, A. (2009) Piecing together a puzzle. An exposition of synthetic biology. *EMBO Rep.*, **10**, 428–432.

Luisi, P.L. (2007) Chemical aspects of synthetic biology. *Chem. Biodivers.*, **4**, 603–621.

O'Malley, M.A., Powell, A., Davies, J.F., and Calvert, J. (2008) Knowledge-making distinctions in synthetic biology. *Bioessays*, **30**, 57–65.

Schmidt, M., Ganguli-Mitra, A., Torgersen, H., Kelle, A., Deplazes, A., and Biller-Andorno, N. (2009) A priority paper for the societal and ethical aspects of synthetic biology. *Syst. Synth. Biol.*, **3**, 3–7.

Sterling, A. (1999) On Science and Precaution In the Management of Technological Risk. An ESTO Project Report Prepared for the European Commission–JRC Institute Prospective Technological Studies. http://ftp.jrc.es/EURdoc/EURdoc/EURdoc/eur19056en.pdf (accessed 7 September 2010).

1
Biofuels

Markus Schmidt, Manuel Porcar, Vincent Schachter, Antoine Danchin, and Ismail Mahmutoglu

1.1
Biofuels in General

1.1.1
Introduction

In 1973, over 86% of the world's total primary energy supply came from fossil fuels. While the energy supply has increased since then (from about 6 Gtoe[1]) in the 1970s to 12 Gtoe in 2007), the share of fossil fuels remains high. In 2007, still over 81% came from fossil fuels (gas: 20.9%; oil: 34%; coal: 26.5%; IEA, 2009). The European Community (EC) is strongly dependent on fossil fuels for its transport needs and is a net importer of crude oil (EC, 2010). Numerous experts predict that oil production will reach a ceiling by 2020, while the demand will continue to grow, pulled by China and India. Facing this demand calls for finding alternatives in petroleum products. At the same time, concerns are increasing about climate change and the potential economic and political impact of limited oil and gas resources. To address these issues and reduce our dependency on fossil fuels the EC has adopted measures[2]) to encourage the production and use of sustainable biofuels (e.g., achieve 5.75% of biofuels among total fuel in the EC). Interestingly, the agricultural policy in Europe or in the United States is also probably the most important driver for the biofuel production. From 2000 to 2008, biofuel production in the United States has increased 82% even though that market accounts for less

1) Gtoe: Giga tonne oil equivalent = 1000 Mtoe (Mega tonne oil equivalent).
2) See, e.g., the Biofuels Directive (2003/30/EC) http://ec.europa.eu/energy/res/legislation/doc/biofuels/en_final.pdf followed by: the Directive 2009/28/EC on the promotion of the use of energy from renewable sources http://eur-lex.europa.eu/LexUriServ/LexUriServ.do?uri=OJ:L:2009:140:0016:0062:EN:PDF; A Strategy for Biofuels http://www.biofuelstp.eu/downloads/An_EU_Strategy_for_Biofuels_2006.pdf; the European Strategic Energy Technology Plan (SET-Plan) http://www.biofuelstp.eu/downloads/SET-PlanCOM_2006_847.pdf; priority is to be given to biofuels research in EC-FP7.

Synthetic Biology: Industrial and Environmental Applications, First Edition. Edited by Markus Schmidt.

than 5% of total fuel consumption. Nevertheless, current biofuel production is based only on the exploitation of the storage organs of agricultural plants (sugars or oils). Research and development efforts are necessary to diversify the feedstock available, to limit the impact on food markets, and to produce more efficient molecules as biofuels. This new SB application is probably one of the success keys, as suggested by new petroleum investments (Exxon's investment in Craig Venter's synthetic genomics start up, BP invested in Qteros, French Total invested in Gevo, Amyris and Coskata).

Biofuels from biomass (e.g., plant stalks, trunks, stems, leaves) are designed to significantly reduce dependence on imported oil and decrease the environmental impacts of energy use. Biotech research is critical for accelerating the deconstruction of (cellulosic) biomass into sugars that can be converted to biofuels. Woodchips, grasses, cornstalks, and other ligno-cellulosic biomass are abundant but more difficult to break down into sugars than cereals (corn, wheat, etc.), a principal source of fuel ethanol production today. Cellulosic ethanol is therefore one of the proposed cornerstones for our energy needs. There are, however, other alternatives both in terms of feedstock and end product (e.g., butanol). Butanol is assumed to hold great promise. Aquatic biomass, such as algae, does not compete with arable land for food production. Algae can be used for the production of a variety of products, including biodiesel and hydrogen. In a long-term perspective, producing hydrogen or even electricity directly from solar energy and water by means of artificial photosynthesis would provide an almost unlimited source of energy (Thomassen *et al.*, 2008).

Biotechnology and especially synthetic biology can play a key role in increasing production and promoting the use of sustainable bioenergy through:

- Development of next-generation biofuel feedstocks,
- Advanced sunlight to biomass to bioenergy conversion,
- By considering socio-economic and environmental challenges when designing technological solutions.

Table 1.1 provides an overview of different biofuels and the technology and feedstock needed to produce them.

1.1.2
Economic Potential

The European Union and the United States have already created an artificial market through energy policies that specify the required rate of incorporation of biofuels in petroleum products. They also support this path through important tax rebates. However, with the arrival of the oil production ceiling and thus the increase in oil prices, a real market will be created and these sectors are likely to be profitable. According to the IEA, 45 million barrels per day could be supplied by biofuels in 2030, making up the deficit in petrol production (see Table 1.2).

World ethanol production for transport fuel tripled between 2000 and 2007 from 17 billion to more than 52 billion liters, while biodiesel expanded 11-fold from less than 1 billion to almost 11 billion liters. Altogether, biofuels provided 1.8% of the

Table 1.1 Overview of different generations of biofuels (UNEP, 2009).

Traditional biofuels	Basic technology	Feedstocks
Solid biofuels[a)]	Traditional use of dried biomass for energy	Fuel wood, dried manure
First-generation biofuels (conventional biofuels)		
Plant oils[b)]	As transport fuel: either adaptation of motors for the use of plant oils; or modification of plant oils to be used in conventional motors	Rapeseed oil, sunflower, other oil plants, waste vegetable oil
	For the generation of electricity and heat in decentralized power or CHP stations	Rapeseed oil, palm oil, jatropha, other oil plants
Biodiesel	Transesterification of oil and fats to provide fatty acid methyl ester (FAME) and use as transport fuel	Europe: rapeseed, sunflower, soya United States: soya, sunflower Canada: soya, rapeseed (canola) South and Central America: soya, palm, jatropha, castor Africa: palm, soya, sunflower, jatropha Asia: palm, soya, rapeseed, sunflower, jatropha
Bioethanol	Fermentation (sugar); hydrolysis and fermentation (starch); use as transport fuel	United States: corn Brazil: sugar cane Other South and Central American countries: sugar cane, cassava Europe: cereals, sugar beets Canada: maize, cereals Asia: sugar cane, cassava Africa: sugar cane, maize
Biogas (CH_4, CO_2, H_2)	Fermentation of biomass used either in decentralized systems or via supply into the gas pipeline system (as purified biomethane): (1) To generate electricity and heat in power or CHP stations (2) As transport fuel, either 100% biogas fuel or blending with natural gas used as fuel	Energy crops (e.g., maize, miscanthus, short rotation wood, multiple cropping systems); biodegradable waste materials, including animal sewage

(Continued)

Table 1.1 (*Continued*)

Traditional biofuels	Basic technology	Feedstocks
Solid biofuels	Densification of biomass by torrefaction or carbonization (charcoal)	Wood, grass cuttings, switchgrass; grains; charcoal, domestic refuse, dried manure
	Residuals and waste for generation of electricity and heat (e.g., industrial wastes in CHP)	
Second-generation biofuels (advanced biofuels)		
Bioethanol	Breakdown of cellulosic biomass in several steps including hydrolysis and finally fermentation to bioethanol	Ligno-cellulosic biomass like stalks of wheat, corn stover and wood; "special energy or biomass" crops (e.g., *Miscanthus*); sugar cane bagasse
Biodiesel and "designer"-biofuels such as bio-hydrogen, bio-methanol, DMF[c], bio-DME[d], mixed alcohols	Gasification of low-moisture biomass (<20% water content) provides "syngas" (with CO, H_2, CH_4, hydrocarbons) from which liquid fuels and base chemicals are derived	Ligno-cellulosic biomass like wood, straw, secondary raw materials like waste plastics
Third-generation biofuels (advanced biofuels)		
Biodiesel, aviation fuels, bioethanol, biobutanol	Bioreactors for ethanol (production can be linked to sequestering carbon dioxide from power plants); transesterification and pyrolysis for biodiesel; other pyrolysis for biodiesel; other future technologies	Marine macro-algae or micro-algae in ponds or bioreactors

a) Traditional use of biomass included for complete overview.
b) Also known as straight vegetable oil. Plant oil used as direct fuel in transport is common in German agriculture with about 838 000 tonnes, mostly rapeseed oil, in 2007, representing 1.4% of total fuel consumption in transport.
c) 2,5-Dimenthylfuran.
d) Dimethyl ether.

Table 1.2 Optimistic scenario of alternative fuel introduction until 2020 in the European Union (Biofuels Platform, 2010).

Year	Biofuels (%)	Natural gas (%)	Hydrogen (%)	Total (%)
2005	2	–	–	2
2010	5.75	2	–	7.75
2015	7	5	2	14
2020	8	10	5	23

Table 1.3 Estimates on future economic indicators in the United States, based on the Renewable Fuel Standard as specified in the US Energy Independence and Security Act of 2007 (BIO-ERA, 2009; EPA, 2010).

	2012	2016	2022	2030
Amount of cellulosic biofuel requirement	1.8 bio l	16 bio l	60 bio l	n/a
Amount of advanced (third-generation) biofuel	7 bio l	27 bio l	79 bio l	n/a
Total renewable fuel requirement	56 bio l	84 bio l	136 bio l	n/a
Direct job creation	29 000	94 000	190 000	400 000
Investments in advanced biofuels processing plants	3.2 bio $	8.5 bio $	12.2 bio $	n/a
Direct economic output from the advanced biofuels industry	5.5 bio $	17.4 bio $	37 bio $	113 bio $

world's transport fuel. Recent estimates indicate a continued high growth. From 2007 to 2008, the share of ethanol in global gasoline-type fuel use was estimated to increase from 3.7 to 5.4%, and the share of biodiesel in global diesel-type fuel use from 0.9 to 1.5%. Currently, the main suppliers for transport biofuels are the United States, Brazil and the European Union. Production in the United States consists mostly of ethanol from corn starch, in Brazil of ethanol from sugar cane, and in the European Union mostly of biodiesel from rapeseed. Investment into biofuel production capacity probably exceeded $4 billion worldwide in 2007 and is growing rapidly. Industry, with government support, is also investing heavily in the development of biofuels. The current biofuel production has been stimulated by biofuel subsides, fuel blending mandates, national interest in energy security, climate change mitigation and rural development programs (see Table 1.3).

International trade in ethanol and biodiesel has been small so far (about three billion liters per year in 2006/07), but is expected to grow rapidly in countries like Brazil, which reached a record-high of about five billion liters of ethanol fuel export in 2008 (Figure 1.1). Predictions forecast that global use of bioethanol and biodiesel will nearly double from 2007 to 2017. Most of this increase will probably be due to biofuel use in the United States, the European Union, Brazil and China. But other countries could also develop towards significant biofuel consumption, among them Indonesia, Australia, Canada, Thailand and the Philippines.

Regarding the global long-term bioenergy potential, estimates depend critically on the underlying assumptions, particularly on the availability of agricultural land for non-food production. Whereas more optimistic assumptions yield a theoretical potential of 200–400 $\times 10^{18}$ J/year or even higher, the most pessimistic scenario estimates 40 $\times 10^{18}$ J/year. More realistic assessments considering environmental

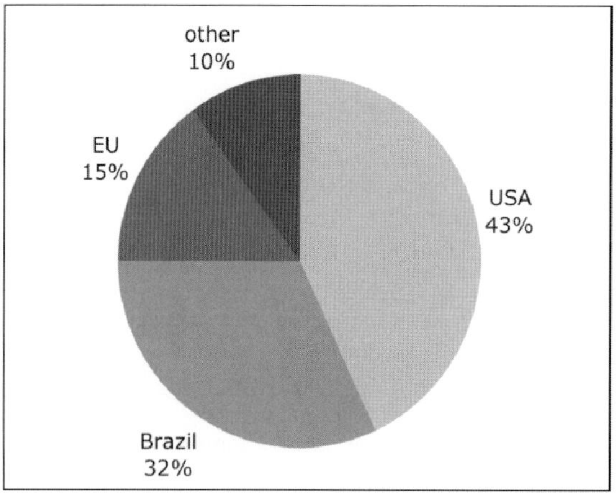

Figure 1.1 Proportion of global production of liquid biofuels in 2007 (Source: UNEP, 2009).

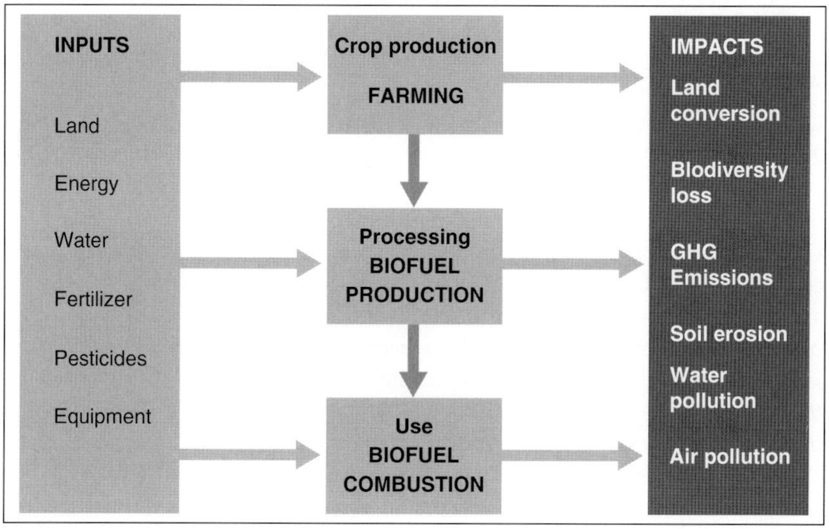

Figure 1.2 Overview of the biofuel life cycle with inputs and relevant environmental impacts.

constraints estimate a sustainable potential of $40–85 \times 10^{18}$ J/year by 2050. For comparison, current fossil energy use totals 388×10^{18} J. These estimations do not take into account biofuel production from non-agricultural biomass (e.g., bio-photovoltaics).

In 2009, total oil and fuel demand worldwide was 84.9 million barrels/day (mb/d), of which 2 mb/day were biofuels (about 2.3%). Not including the United States and Brazil, the global biofuel supply is only about 0.5 mb/day. Examining the global supply of fuel for 2010 reveals only little growth compared to the previous year (Figure 1.2). Among the key sources of global growth in 2010, however, are biofuels (+210 kb/day) (IEA, 2010).

1.1.3
Environmental Impact

The effects of biofuels on the environment include changes in the emission of greenhouse gases (GHG), the displacement of non-renewable fuels, a positive effect on water and air quality, as well as a beneficial influence on biodiversity, eutrophication and acidification of soils. Currently, biofuel production is mainly related to ethanol and biodiesel. These require an agricultural bio-resource such as sugar cane, corn or rapeseed, as well as a considerable amount of water and fertilisers, not to mention pesticides. Today, most environmental effects are strongly linked to the agricultural practices of the bioenergy crop cultivation.

The extension of cropland for biofuel production is continuing, in particular in tropical countries where natural conditions favor high yields. In Brazil, the planted area of sugar cane comprised nine million hectares (Mha) in 2008 (up 27% since 2007). Currently, the total arable land of Brazil covers about 60 Mha. Also, the total cropping area for soybeans, which is increasingly being used for biodiesel, could potentially be increased from 23 Mha in 2005 to about 100 Mha. Such expansion is most likely on the pastureland and in the savannah (cerrado). For example, palm oil expansion in Southeast Asia is considered as the leading cause of the lost biodiversity of rainforest. Taking an example from Indonesia, plantations of 20 Mha for palm oil trees are planned, while the existing stock has already been at least 6 Mha. Two-thirds of these expanded palm oil cultivations in Indonesia will be grown on lands converted from rainforests, with one-third on previously cultivated or currently fallow land. Of the converted rainforest areas, one-quarter contained peat soil with a high carbon content – resulting in particularly high GHG emissions when drained for oil palms. Based on the 2009 estimate by Bringezu *et al.*, by 2030, a share of 50% from peat soils is expected. If current trends continue, in 2030 the total rainforest area of Indonesia will be reduced by 29% as compared to 2005, and will only cover about 49% of its original area in 1990. Figure 1.3 shows the estimated positive and negative environmental effects of different uses of biomass, including liquid transport fuels (Bringezu *et al.*, 2009).

The Scientific Committee on Problems of the Environment (SCOPE, 2009) conducted a project on the environmental impact of biofuels and recommended – in line with Tilman *et al.* (2009) – that: many of the adverse effects of biofuels on the environment (e.g., loss of biodiversity, eutrophication, acidification of soils, water pollution, production of GHG, in particular nitrous oxide) could be reduced by using best agricultural management practices, if production is kept below sustainable production limits.

In general, biofuels made from organic waste are environmentally more benign than those from energy crops. Using biomass primarily for material purposes, reusing and recycling it, and then recovering its energy content can yield multiple dividends (Figure 1.4).

Low-input cultivation of perennial plants, for example, from short-rotation forestry and grasslands, may be an effective source of cellulosic biomass and provide environmental benefits (reduced pollution and lower greenhouse gas emissions).

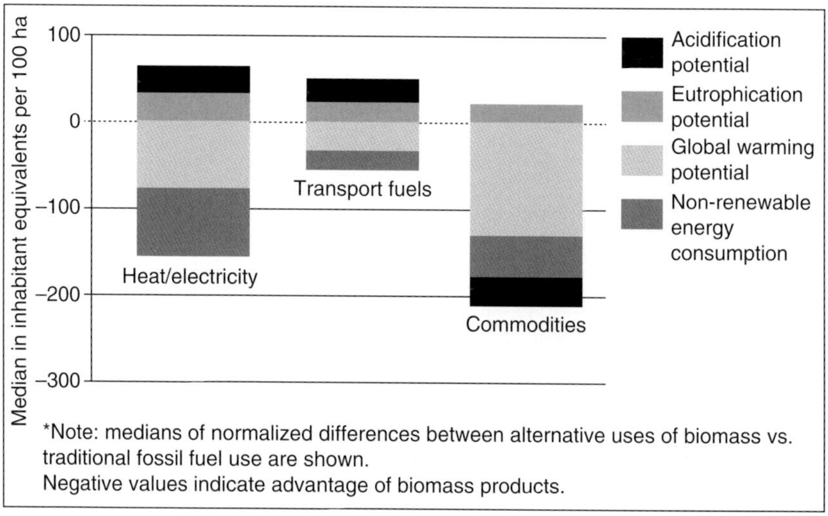

Figure 1.3 Comparative relief of environmental pressure through use of biomass for heat/electricity, transport fuel and material products (Source: UNEP, 2009).

Careful attention to maintaining the long-term productivity of these systems through nutrient additions (particularly potassium) is required.

New liquid hydrocarbon fuels (see Section 1.3) produced from cellulosic biomass are under development and seem likely to offer several advantages over producing ethanol from cellulose in terms of more efficient yields and less environmental impact (Figure 1.5). The economic viability of this technology still needs to be proven, and potential conflicts with traditional wood-based industries should be considered. Note, however, that the aromatics present in lignin provide a good substrate as a traditional fuel, while they are, with our present knowledge, difficult to convert to biofuels.

1.1.3.1 Land Requirements for Projected Biofuel Use

Estimates of land requirements for future biofuels vary widely and depend on the basic assumptions made – mainly the type of feedstock, geographical location, and level of input and yield increase. More conservative trajectories project a moderate increase in biofuel production and use. They have been developed as reference cases under the assumption that no additional policies would be introduced to further stimulate demand. These range between 35 Mha and 166 Mha in 2020. There are various estimates of potentials of biofuel production which calculate cropland requirements between 53 Mha in 2030 and 1668 Mha in 2050. About 118–508 Mha would be required to provide 10% of the global transport fuel demand with first-generation biofuels in 2030 (this would equal 8–36% of current cropland).

The analysis of land availability for an aggressive biofuels program is summarized in the following five points (SCOPE, 2009):

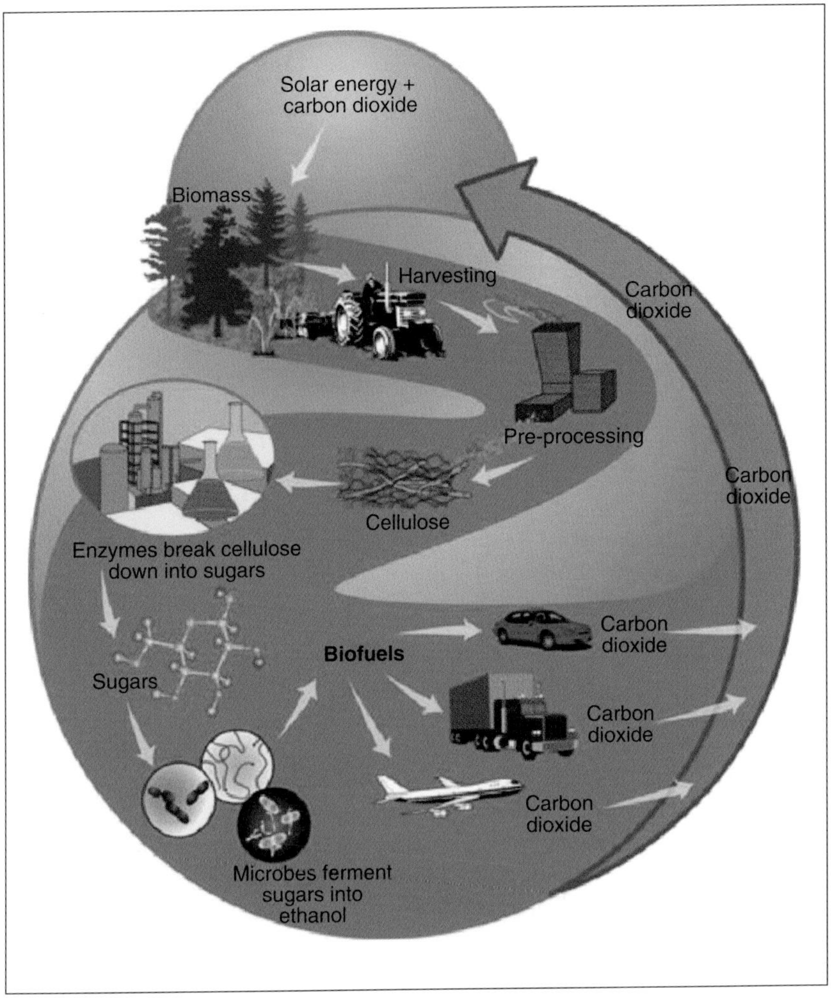

Figure 1.4 Schematic view of the biofuel conversion process (US DOE, 2010).

1) Supply of land is tight and a growing population will put increasing pressures on its uses.
2) How much land is available, at which yield potential, and in which locations to produce enough biofuels to provide a significant fraction of world energy is the subject of much debate.
3) The real pressure points are in the tropics, where new croplands could be developed, where biodiversity values are high, and where much of the population is vulnerable to multiple stresses.

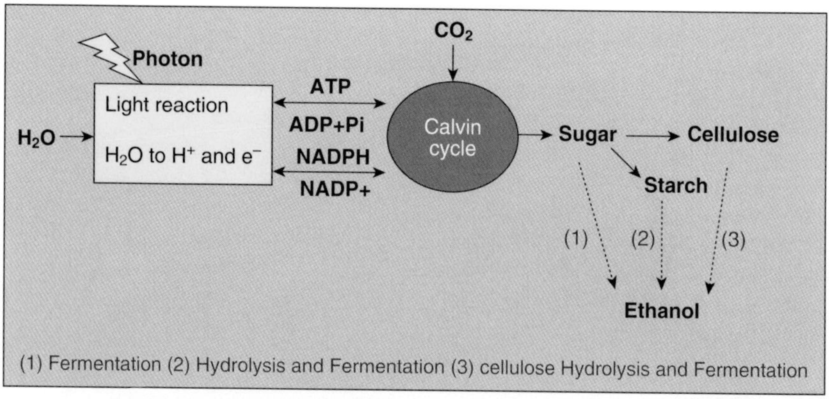

Figure 1.5 Schematic pathway from photosynthesis in plants to biomass to ethanol.

4) From an environmental standpoint, there are few areas where biofuels are an acceptable use of land given the alternative uses.

5) At the regional and local scales there are opportunities to create acceptable uses of biofuels that have net benefits for society.

1.1.3.2 Other Environmental Concerns

It has been suggested that each biofuel should be evaluated based on its net benefit, that is, a full analysis of its effects on net energy gain, the global food system, greenhouse gas emissions, soil structure and soil fertility, water and air quality and tradeoffs of biodiversity (Tilman *et al.*, 2009). David Tilman has proposed that biofuels should be derived from feedstocks produced from perennial plants grown on degraded lands, crops residues, sustainably harvested wood and forest residues, double crops and mixed cropping systems, as well as municipal and industrial waste. The biofuels derived from those feedstocks will maximize their benefits by not competing with food crops, minimizing impacts on land clearing and offering real GHG reductions.

1.1.3.3 Impact of Legislative Decicions

In the Biofuels Directive (EC, 2003) the EC mentioned a concrete target for the share of biofuels: "*Member States should ensure that a minimum proportion of biofuels and other renewable fuels is placed on their markets, and, to that effect, shall set national indicative targets. A reference value for these targets shall be 5.75%, calculated on the basis of energy content, of all petrol and diesel for transport purposes placed on their markets by 31 December 2010*". Setting such a target, although with good intentions, may alsol cause unintended consequences. Some observers argued that these goals could only be met at an enormous environmental cost, including higher global CO_2 emissions, massive biodiversity loss, and pollution of soil and water in biofuel production countries (Green, 2010). To cope with rising concerns of unwanted side effects of biofuels, some countries – including the EC – have started to promote criteria for sustainable bioenergy production. The EC decided to adapt their targets for biofuels, and specify under which conditions these targets should be met. The Directive 2009/28/EC on the promotion of the use of energy from renewable sources (EC, 2009) sought to correct the potential incentives for unsustainable developments in order to meet the requirements laid out in the 2003 directive. Confronted with the risk that imports to the EU from biomass- and biodiversity-rich countries like Brazil could actually promote the deforestation of rainforests, the 2009 directive (EC, 2009) formulated clear priorities for the production of biofuels. In its Article 17 on "Sustainability criteria for biofuels and bioliquids" it states for example: "*Biofuels and bioliquids. . . . shall not be made from raw material obtained from land with high biodiversity value . . .*"

This example shows that solving one environmental problem (replacing fossil with renewable fuels) may well lead to unintended second-order problems in another area (e.g., deforestation of rainforest to make land for energy crops). This calls for ensuring that the production of biofuels and other forms of bioenergy is carried out in a sustainable manner (See also Section 3.5 on cellulosomes).

1.1.4
Foreseeable Social and Ethical Aspects

Large-scale production of biofuels and feedstocks will require considerable infrastructure to transport the biomass used upstream of the processes. It may also lead to a further concentration of landholdings and transformation of the rural landscape. In contrast, small-scale production of biofuels can provide local energy security or access and, if managed properly, can have no adverse impacts on food production. If development programmes select small communities for the local production of electricity using biofuels, intra-country inequalities can be reduced. Europe has examples of such small-scale production (such as wood granulates) that are sustainable in both social and environmental terms. These could be adapted for the development of similar programmes using varied feedstocks and management practices in communities in Africa, Asia and Latin America. For example, eco-friendly energy farms have been promoted in Norway and other Scandinavian countries, where small farms produce their own energy (mostly heat and biodiesel) by using biofuels produced locally (Figure 1.6). This model may be applied in local communities in certain developing countries to satisfy local energy needs (SCOPE, 2009).

Producing biofuel feedstocks may promote rural development in some traditional agriculture countries, but the social impacts need to be assessed carefully. For instance, increased demand for palm oil for biodiesel provides increased employment in countries like Indonesia. At the same time, negative social impacts have already been reported. These include poor wages, low labor standards, impact on health and local culture, land grabbing and loss of environmental goods (Tilman *et al.*, 2009). Whether or not second- or third-generation biofuels will bring higher social benefit depends mainly on the economically feasible size of the production

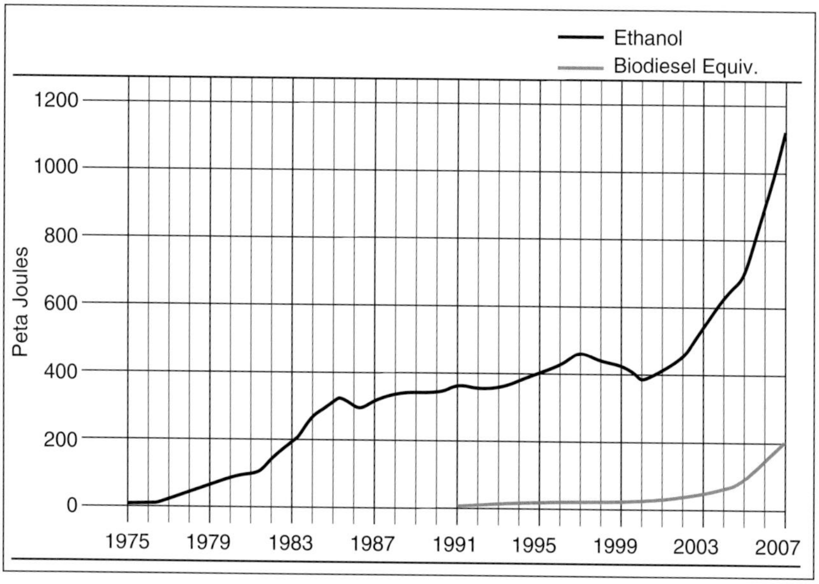

Figure 1.6 Global bioethanol (black line) and biodiesel (gray line) production 1975–2007 (source: UNEP, 2009).

unit. The larger the size of the agricultural field, factory or tank, the higher the probability that only big business will be able to benefit first hand from biofuel production. Future technology developments should consider the design of small- and medium-scale units that are more accessible to SMEs in order to allow for a more socially distributed benefit.

1.1.4.1 How Could the New SB Application Impact Society at Large?

Biofuels and food. The production of biofuels, mainly first-generation (simple sugar- and starch-based) bioethanol, has affected the price of food in the past. Those biofuels that require significant land for their production and/or compete with crops can have this effect. This negative effect is discussed in detail in Section 1.2.

Greenhouse effect. Burning biofuels emits CO_2 and thus theoretically contributes to global warming. This emission is substantially compensated by the succeeding crop, which will absorb atmospheric carbon. Note, however, that transportation, agricultural practices and use of fertilisers will emit additional CO_2. Thus, the net emission of CO_2 is not zero, but it is certainly much lower than that of the fossil fuels.

Socio-ecological impact. The potential side-effects of massive conversion of land surfaces for fuel production include: displacement of traditional cultures, competition over arable land, and indirect contribution to global warming or desertification. This is more a political than a scientific problem because the legal framework and the sustainability of the use of the technology are central. A case by case scenario is described in the following issues.

In general, social interactions may change due to major shifts between traditional farming and biofuel production. Even in those cases where the substrates are not crops (i.e., algal production), competition for space might occur.

Ethical issues. These depend on the biofuel. Factors such as toxicity, flammability or ease of local production may vary considerably. These topics are discussed in the following chapters

Justice of distribution. This is a key point in the ethics of biofuels. It is difficult to forecast the consequences because the future economy might rely on one, several or a mix of biofuels. The kind of biofuel can also have important consequences: some can be either extremely easy to produce *in situ* for local consumption (i.e., vegetable oil for direct use in old cars), relatively easy to produce but difficult to manipulate (hydrogen) or process (algal biodisel), or show intermediate features (bioethanol). Justice of distribution will likely depend on:

- The nature of the biofuel itself (ease of production and manipulation, security).
- Intellectual property rights, linked to the complexity of the technical system set in place, which might lead to monopolies.
- Political will.

1.2 Ethanol

1.2.1 Introduction

Ethanol fermentated directly from starch or sugar, as a typical first-generation biofuel, is the most common liquid biofuel used today (Figure 1.7). (Ligno)cellulosic ethanol, an advanced biofuel, can be produced from a variety of biomass materials. The standard procedure includes:

- A "pretreatment" phase, to make the cellulosic material amenable to hydrolysis.
- Cellulose hydrolysis, to break down the cellulose molecules into sugar molecules, commonly with *Trichoderma reesei* enzymes.
- Separation of the sugar solution from the residual materials (e.g., lignin).
- Microbial fermentation of the sugar solution, usually by the budding yeast *Saccharomyces cerevisiae*.
- Distillation and dehydration to meet biofuel standards (concentration >99.5%).

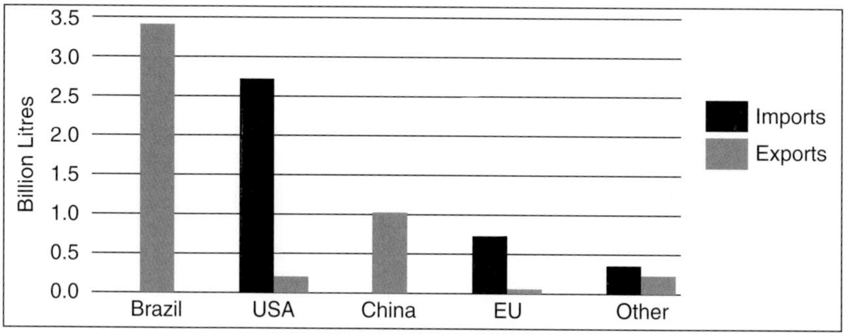

Figure 1.7 International trade in ethanol in 2006 (source: UNEP, 2009).

This conventional process for the production of bioethanol from lignocellulosic biomass is often called separate hydrolysis and fermentation (SHF); modifications of the procedure involving simultaneous saccharification and fermentation (SSF) are also used. In SSF, hydrolytic enzymes and yeast are incubated together under compromised conditions, and ethanol is obtained in one step. The main advantage of this procedure is the lack of end-product inhibition. The main disadvantage is that the compromise conditions are usually suboptimal for both hydrolysis and fermentation. Bioethanol can be produced from biomass such as sugarcane and corn, but lignocellulosic material (i.e., rice straw, wood, paper waste) has an incredible potential for bioethanol production due to the enormous amount of such biomass produced worldwide every year. Assuming that about 33% of the dry weight of the straw is cellulose, and considering that the theoretical ethanol yield is 0.511 g of ethanol per gram of glucose, 1 kg of rice straw could yield 168 g of ethanol. This means that as much as 134 million tonnes of bioethanol would be produced if all the rice straw on earth were converted into biofuel. The efficient transformation of lignocellulosic biomass to ethanol is a challenge for SB, and this approach is expected to have a major influence on the production of bioethanol (see Table 1.4).

1.2.2
Economic Potential

Currently, only the ethanol from sugarcane in Brazil is economically viable. The corn-derived ethanol in the United States and the beet-derived ethanol in France are only feasible because they are supported by government subsidies. For the cellulosic ethanol, the "90-billion gallon biofuel deployment study" by the Sandia National Laboratories and the General Motors R&D center showed that cellulosic ethanol can compete with oil at $90/bbl based on the following assumptions:

- Average conversion yield of 95 gallons per dry ton of biomass;
- Average conversion plant capital expenditure of $3.50 per installed gallon of nameplate capacity;
- Average farm-gate feedstock cost of $40 per dry ton.

Table 1.4 Evolution of bioethanol production in selected European Union countries.

Country	Annual production (million liters/year)						
	2002	2003	2004	2005	2006	2007	2008
France	114	103	101	144	293	539	950
Germany	0	0	25	165	431	394	581
Spain	222	201	254	303	402	348	346
Poland	63	76	48	64	120	155	200
Hungary	0	0	0	35	34	30	150
Slovakia	0	0	0	0	0	30	94
Austria	0	0	0	0	0	15	89
Sweden	63	65	71	153	140	120	76
Czech Republic	6	0	0	0	15	33	76
United Kingdom	0	0	0	0	0	20	75
Other EU countries	0	0	29	49	173	119	216

These simulations assume technological progress in the conversion technologies in which SB plays a central role. The cost competitiveness of ethanol depends directly on the price of oil and the realization of technological improvements.

A major problem in the large-scale use of ethanol is that it needs to be transported (as in the case of oil). Yet it is an extremely corrosive material, and most of the existing infrastructure used to transport oil would have to be replaced if ethanol played an important role. Transport using trucks, for example, would considerably decrease the interest in its use. Furthermore, its hygroscopy makes it a very awkward fuel for sensitive engines, such as those used in airplanes.

1.2.3 Environmental Impact

Current life cycle assessments (LCA) of biofuels show a wide range of net greenhouse gas savings compared to fossil fuels. This mainly depends on the feedstock and conversion technology, but also on other factors such as methodological assumptions. For ethanol, the highest GHG savings are recorded for sugarcane (70% to more than 100%), whereas corn can save up to 60% but may also cause 5% more GHG emissions (see Table 1.5). (The highest variations are observed for biodiesel from palm oil and soya.) The high savings of the former depend on high yields, those of the latter rely on the credits of byproducts. Negative GHG savings, that is, increased emissions, may result in particular when production takes place on converted natural land and the associated mobilization of carbon stocks is accounted for (Figure 1.8). Besides GHG emissions, other impacts of biofuels, such as eutrophication, are also relevant and contribute to significantly worsened environmental quality in certain regions of the world.

Most of the currently used crops for transport biofuels are also food crops. Global land use for the production of fuel crops recently covered about 2% of global

Table 1.5 Lifecycle GHG thresholds specified in US Energy Independence and Security Act (EISA; % reduction from 2005 baseline; EPA, 2010).

Type of fuel	GHG savings (%)
Renewable fuel	20
Advanced biofuel	50
Biomass-based diesel	50
Cellulosic biofuel	60

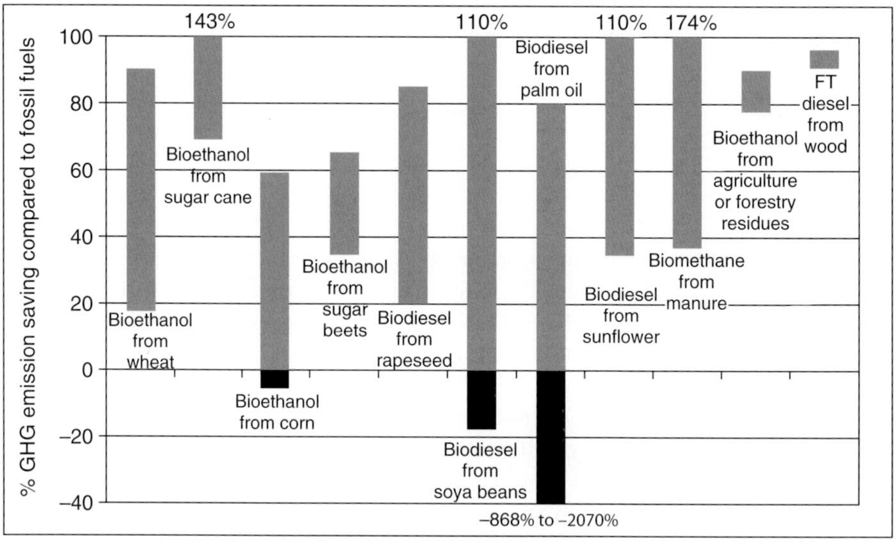

Figure 1.8 Green-house gas (GHG) savings from bioethanol (and biodiesel) compared to fossil fuels (Source: UNEP, 2009).

cropland (about 36 Mha in 2008). This development is driven by volume targets rather than by land use planning. The extension of cropland for biofuel production is continuing, in particular in tropical countries where natural conditions favor high yields. In Brazil, the planted area of sugarcane comprised nine million hectares in 2008 (up 27% since 2007). Currently, the total arable land of Brazil covers about 60 Mha.

It is difficult to calculate GHGs from biofuels because the production systems are inherently complex and the methods used to quantify savings are subjective. Several factors affect emissions, among them emissions embodied in biomass production, the use of electricity in the conversion process, indirect land-use change and fertilizer replacement (Figure 1.9). Indirect emissions from biofuels should also be taken into account; this occurs when biofuel production on agricultural land displaces agricultural production. This type of land-use changes will increase net GHG emissions (Melillo *et al.*, 2009).

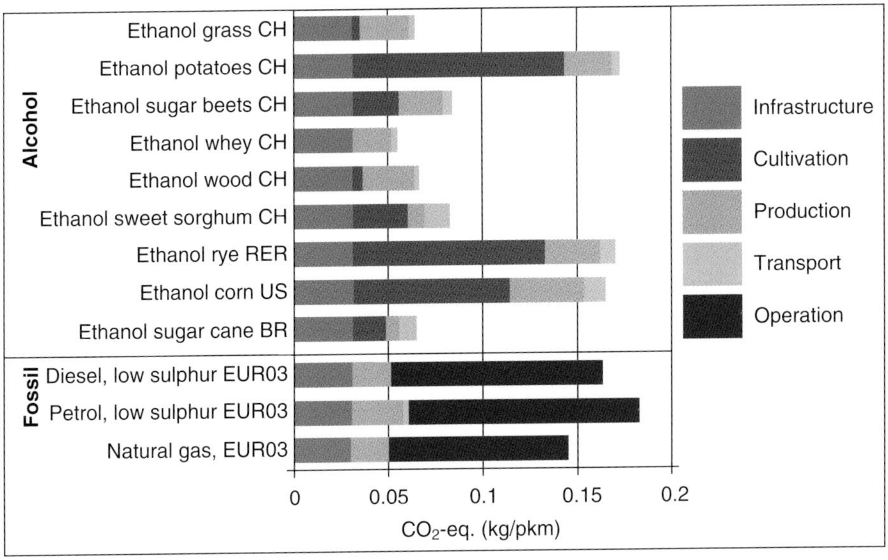

Figure 1.9 Comparison between the greenhouse gas emissions of bioethanol for transport and fossil fuels, differentiated by phase. Ethanol production clearly does not automatically lead to a reduction in GHG emissions. (The unit is kilogram CO_2 equivalent per person kilometer; Zha, 2007; UNEP, 2009).

Table 1.6 Companies using synthetic biology to produce biofuel.

Companies	Products
Amyris[a)]	Renewable diesel
Dyadic International	Ethanol, enzymes
GEVO	Butanol, isobutanol, chemicals
Joule Biot	Ethanol, chemicals
LS9	Renewable gasoline, diesel, jet fuel, chemicals
Metabolic Explorer	Butanol, 1,3,propanediol
OPX Biot	Renewable acrylic
Solazyme	Green crude, biodiesel, algal oil, others
Synthetic Genomics	Algal oil, biodiesel, chemicals
Vendezyne	Ethanol

a) Amyris left the biofuel sector for lack of positive results in early 2012.

The carbon neutral hydrocarbons (CNHCs) have been brought into play to define those biofuels compatible with the existing transportation infrastructures and therefore capable of a gradual deployment with minimum supply disruptions (Zeman and Keith, 2008). Ethanol is seen by many as such a CNHC, although its real environmental benefits could be rather limited compared to alternative next-generation biofuels (see Table 1.6).

Because of the environmental (and social) impact of big agribusiness, many researchers have recommended that instead of cultivating energy crops for ethanol, a more environmentally friendly biomass to biofuel conversion can be achieved by using approaches (Tilman *et al.*, 2009) such as:

- Perennial plants grown on degraded lands abandoned by agricultural use;
- Crop residues;
- Sustainably harvested wood and forest residues;
- Double crops and mixed cropping systems;
- Municipal and industrial wastes.

1.2.4
Foreseeable Social and Ethical Aspects

Current large-scale (ethanol) first-generation biofuel production in Brazil (and to a lesser extent other countries) has done little to improve the economic situation of the average fuel consumer or promote sustainable rural development to the benefit of small-holder farmers or the urban poor (Rodrigues and Moraes, 2007; Sawyer, 2008; SCOPE, 2009). Little evidence and only few arguments suggest that future large-scale biofuel production will have a positive social effect (ETC, 2008). Indeed, the small body of research on the socio-economic impacts of biofuels suggests that an expansion of the areas under feedstock cultivation for biofuels will benefit only large land owners, speculators and urban elites in developing countries. Small-scale farmers and the poor in developing countries will be negatively impacted. In contrast, very small-scale biofuel production for local consumption shows some promise in the provision of energy security and improvements in human well-being and equity (EEA, 2006).

An onging debate, especially during the food crisis in 2008, was about the ramifications of cultivating energy crops *instead* of food crops. The following quote shows the relevance of the problem:

> *"And what I've said is top priority is making sure people are able to get enough to eat. If it turns out we need to make changes in our ethanol policy to help people get something to eat, that has got to be the step we take . . . We have rising food prices around the United States. In other countries, we're seeing riots because of the lack of food supply, so this is something we're going to have to deal with."*
>
> Barack Obama (D-IL) on Meet the Press, Sunday, May 4, 2008

The successful application of SB to bioethanol production might, as for biofuels in general, create a new business with the corresponding social classes (producers, traders, sellers, consumers, etc). The nature of the social interactions is not expected to differ from that of any other new and successful energy-related business.

If SB-based strategies succeed, the volume of ethanol produced each year might dramatically increase. For example, the calculations mentioned above on straw-based bioethanol show that a huge volume corresponding to eightfold the worldwide production of bioethanol in 2008 might be achieved. And this scenario considers only this particular lignocellulosic biomass. The implementation of procedures to efficiently produce bioethanol from molasses might also be a reality in the near future (Hatano *et al.*, 2009). If these predictions are met, the impact on society will be deep. Although the societal consequences of a dramatic turnover in the use and availability of biofuels are difficult to forecast, at least these major issues arise:

Stability of ethanol as a major biofuel: the price factor. The price of the ethanol might suffer from unpredictable variations linked to the market economy. This would be the case if bioethanol production increased in such a way that a significant substitution of fossil fuels occurs. It would also be the case in a scenario in which bioethanol dominates the biofuels market. The impact of these variations would affect both producing countries (see last issue) and consumer countries. It seems reasonable to forecast that, in a future economy based on biofuels, a parallelism with the situation of oil can occur. The price of a barrel of oil is difficult to forecast over the long term. Likewise, ethanol might be subject to price fluctuations that would affect its competitiveness with other fuels.

Price of food. The effect of massive bioethanol production associated with advances in the SB approaches designed to take advantage of lignocellulosic biomass should not directly effect food prices because byproducts rather than grains or vegetables would be used to produce ethanol. Nonetheless, food prices might well rise in scenarios in which significant portions of the surface dedicated to basic crops were to be transformed into ethanol-producing varieties. It is important to highlight that, rather than being a scientific issue, the use of a novel technology is the responsibility of policy makers. This calls for a legal framework that fosters the use of agricultural lignocellulosic byproducts such as prune rests or straw, and that strongly regulates the transformation of food-producing crops.

Greenhouse effect. The impact of a bioethanol-based economy on reigning in global warming is expected to be high. Indeed, the conversion of lignocellulosic biomass into biofuels represents huge CO_2 savings: although CO_2 is released when ethanol is burned, this emission, unlike that of fossil fuels, is of recent atmospheric origin and it is fixed again by the following harvest. The result: the net CO_2 emission of bioethanol is close to zero. Nonetheless, a regulatory framework limiting the artificial extension of the crops for ethanol production is envisaged. This is because certain cultures that yield more lignocellulosic biomass, such as paddy rice, are net greenhouse gas producers: paddy culturing is in fact thought to be responsible for global methane emissions of as much as 28.2 Tg/year.

1.2.4.1 Could the Application Change Social Interactions?

The successful application of SB to bioethanol production might, as for biofuels in general, create a new business with the corresponding social classes (producers, traders, sellers, consumers, etc). The nature of the social interactions is not expected to differ from that of any other new and successful energy-related business.

Ethical issues of applying SB to improve ethanol production for biofuel use involve:

i) Ethics of the technology;
ii) Ethics of the product of the technology;
iii) Ethics of the use of the product (bioethanol).

Regarding (i), no major differences are evident compared to standard biotechnology such as transgenic crops or GM microorganisms expressing enzymes for the textile industry. On (ii), ethanol itself is a well-known chemical compound with a rather low toxicity and very well-characterized effects. The inflammability of ethanol and its transportation might, by contrast, be a source of conflict of interest and security hazards. These, along with the ecological or economic issues described here, correspond to the application (iii) of the technology rather than the technology itself.

1.2.4.2 Producing Countries, Rich Countries?

If the current scenario of first-generation bioethanol is maintained in the future, then countries such as Brazil or the United States will probably monopolize the first-generation bioethanol production as they do now. However, significant advances in processing lignocellulosic biomass might change the world distribution of producers and consumers. Indeed, lignocellulosic biomass is present evenly wherever vegetables grow, that is, worldwide. Nonetheless, tropical and sub-tropical regions along with certain moderate climate areas have a larger production potential because of the climate. The expectation therefore is that the production of lignocellulosic bioethanol would be mainly based in countries with the ability to produce grain, such as wheat, rice and maize. South-Eastern Asia, North America, and to a lesser extent, Europe would then become the major producing regions (see Table 1.7; Domínguez-Escribà and Porcar, 2010).

Access to Bioethanol Depending on the type of technology developed to transform biomass into bioethanol, the required steps might be performed in large, confined structures. In a contrasting scenario, local producers would be involved. If the latter was the case, the farmers of remote areas would be autonomous in terms of fuel production and use. Alternatively, if the technology involved requires (or is anyhow monopolized by) large biochemical enterprises, the availability and thus the price of the final product will be controlled by the manufacturers. Beyond the limitations of the technology (enzymatic reactions might be difficult to control by local producers, for example), political will is a key point to ensure fair access to bioethanol.

Table 1.7 Consumption of vehicle fuels in Europe (EU-27; Eurostat, 2005[a]).

Country	Consumption of vehicle fuels in Europe (PJ)				Objectives 2010
	Gasoline	Diesel	Biofuels	Total	Biofuels
Belgium	2410	7298	0	9846	665
Germany	31379	29522	2673	63720	4570
Spain	9928	27351	357	37718	2571
France	14322	35254	597	50479	3447
Italy	18397	26539	224	47470	3256
Netherlands	5603	7370	–	13624	1022
Poland	5392	6313	76	14507	947
Sweden	5281	3539	289	9110	680
United Kingdom	25615	22898	111	48845	3655
EU-27	148144	203828	4413	364494	25603

a) See: epp.eurostat.ec.europa.eu

1.3 Non-ethanol Fuels

1.3.1 Introduction

Ethanol is considered by many researchers and engineers as a sub-optimal biofuel due to technical challenges (corrosive and extremely hygroscopic: it tends spontaneously to be associated to 4% water). Ethanol is not the best energy molecule and poses many problems of transport and incorporation in gasoline. In addition, as the European vehicle market is very focused on diesel engines, this molecule represents only a moderate interest there. Furthermore, the power energy of this molecule is insufficient to provide the energy needed to fly an airplane.

In terms of energy yield, the fact that ethanol is already a partially oxidized derivative of carbon does not make it a fuel of choice. Nonetheless, ethanol is not the only biofuel available; a number of different chemical substrates have been investigated as alternative liquid transportation fuels, for example, fatty acids, fatty alcohols, biodiesel, biobutanol, biopropanol, acetone, methanol. For comparison, Figure 1.10 shows the energy densities of a variety of fuel substrates.

Biodiesel is commonly produced by the transesterification of vegetable and animal oils. It is composed of fatty acid methyl and ethyl esters. Based on the current techniques and lifecycle accounting, biodiesels from soybean yield more net energy gain (93%) than ethanol from corn grains (25%) (Hill *et al.*, 2006). However, the problem of availability of traditional agricultural resources and the European diesel demand led some companies to develop alternative production pathways: companies like Amyris and LS9 have used SB technologies to

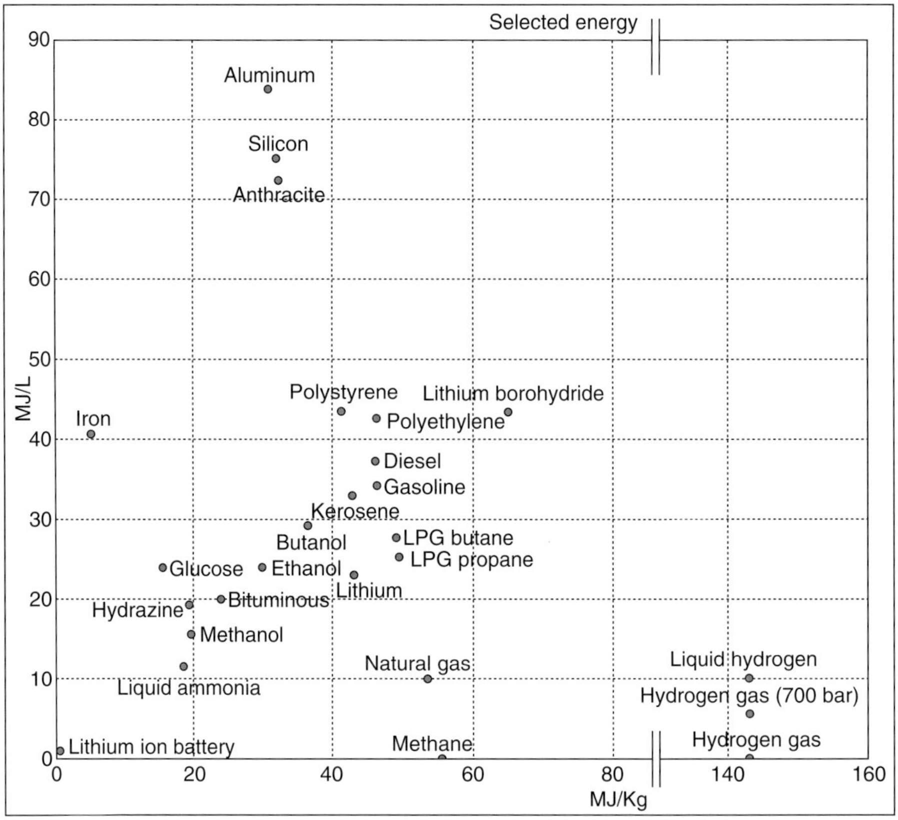

Figure 1.10 Energy densities for selected fuel substrates.

develop the new pathway of sugar to diesel. Amyris, based on the work of Jay Keasling on the isoprenoid pathway, wanted to create so-called truly "No Compromise® fuels". Amyris wrote: "these renewable fuels don't demand sacrifice in performance or penalty in price, and offer a superior environmental profile by reducing lifecycle (GHG) emissions of 80% or more compared to petroleum fuels. They are also delivered to consumers using existing petroleum distribution infrastructure and work in today's engines." Amyris stopped its biofuel R&D in early 2012.

Fatty acids are the raw material for producing biodiesel (renewable diesel) and other products. Fatty acids react like any other carboxylic acid, which means they can undergo esterification and acid-base re-actions. Reduction of fatty acids yields **fatty alcohols**, which can be used as biofuels (Lestari *et al.*, 2009; Steen *et al.*, 2010). To yield a better fuel, biodiesel could be produced using bio-ethanol for the trans-esterification to produce ethyl esters, which are less viscous (Kleinov *et al.*, 2007). A possible longer-term route to biodiesel, currently under active investigation, involves the direct production of fatty acid ethyl esters from lignocellulose by engineered organisms such as Actinomycetes and also via production of wax esters in plants (Kalscheuer, Stolting, and Steinbüchel, 2006; Steen *et al.*, 2010). The future of this pathway is linked to two key challenges: cost and scale. Currently, this is not competitive with ethanol production. Nevertheless, the producers claim that costs will decrease by commercialization and point out that gasoline is far more lucrative than ethanol.

Butanol. Third-generation biofuels include alcohols like **bio-propanol** or **bio-butanol**, which due to current lack of production experience are usually not considered to be relevant as fuels on the market before 2050 (OECD/IEA, 2008). Increased investment could accelerate their development. The same feedstocks as for first-generation ethanol can be used, but requires more sophisticated technology. Propanol can be derived from chemical processing such as dehydration followed by hydrogenation. As a transport fuel, butanol has properties closer to gasoline than bioethanol (UNEP, 2009). Butanol is generally considered a better alternative fuel than ethanol. As a "higher" (4-carbon) alcohol, it more closely resembles the hydrocarbons in gasoline (usually 4–12 carbon atoms) and in diesel (usually 9–23 carbon atoms).

Compared with ethanol, butanol has higher energy density, can be transported by the current infrastructure of pipelines, and can be used directly in **conventional petrol engines**. Butanol has been produced along with acetone and ethanol in the Acetone butanol ethanol (ABE) fermentation by *Clostridium acetobutylicum*, with the drawback of low yields and **low product concentrations** due to the toxicity of butanol to the fermentative strains (Lee *et al.*, 2008). To enhance the yield, the butanol production pathway has been engineered in *Escherichia coli* (Figure 1.11; Atsumi and Liao, 2008a, 2008b). An even higher productivity was achieved by introducing the isobutanol pathway while also overexpressing a key enzyme in the photosynthetic pathway in *Synechococcus elongatus* (Atsumi, Higashide, and Liao, 2009). Alcohols with longer carbon chains would be even better (e.g., octanol). It is clear, however, that alkanes would always be preferred choices whenever possible. The future of this pathway is linked to two key challenges: (i) demonstrating that there is a process with which the butanol can be commercially and economically produced at scale; and (ii) market adoption.

Acetone. Compared with other alcohols, acetone, one of the co-products from the ABE fermentation, can be recovered easily by direct exhaust from the gas stream due to its high volatility. It can be produced in engineered *E. coli* (Bermejo, Welker, and Papoutsakis, 1998). It can be reduced to isopropanol, another potential biofuel or chemical feedstock, from strains co-expressing a secondary alcohol dehydrogenase (Hanai, Atsumi, and Jiao, 2007).

Methanol can be produced from a wide range of biomass feedstocks via a thermochemical route similar to the Fisher–Tropsch process for BtL, a process creating synthetic fuels made from biomass through a thermochemical route. It can be blended in petrol at 10–20%. Methanol can be converted to dimethylether

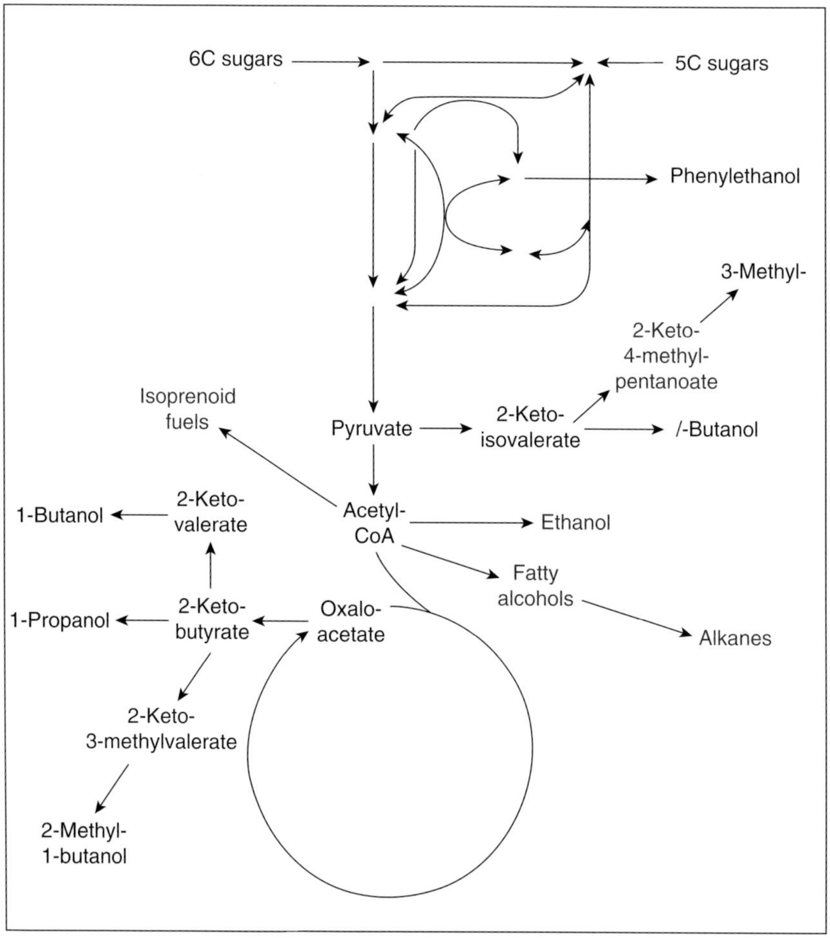

Figure 1.11 Metabolic engineering routes in *E. coli* to deliver next-generation biofuels (Source: Keasling and Chou, 2009).

(DME) by catalytic dehydration. Above –25 °C or below 5 bar, DME is a gas. Hence its use as a transport fuel is similar to that of liquefied petroleum gas (LPG). It cannot be blended with standard diesel fuels. DME can also be created directly from syngas. The BioDME project[3] aims to demonstrate production of environmentally optimized synthetic biofuel from lignocellulosic biomass at an industrial scale. The Supermethanol project on reforming crude glycerine in supercritical water to produce methanol for re-use in biodiesel plants (FP7-212180)[4] aims to produce methanol from crude glycerine, and then re-use the

3) http://www.biodme.eu/about-dme
4) http://www.supermethanol.eu/

methanol in the biodiesel plant. This will improve the energy balance, carbon performance, sustainability and overall economics of biodiesel production. Glycerine is a co-product of the biodiesel process. The work expands on expertise generated by the consortium on reforming glycerine in supercritical water, and then producing a synthesis gas suitable for direct once-through methanol synthesis (glycerine to methanol; GtM). Producers will be less dependent on the methanol spot price: there is a (partial) security of methanol supply, and their byproduct is used as a green, sustainable feedstock.

Isobutene. The French company Global Bioenergies developed a totally artificial metabolic pathway to produce the gaseous (predecesor) biofuel isobutene.[5] The metabolic pathway developed by Global Bioenergies makes it possible to transform renewable resources, such as sugar cane, beet or cereals, into isobutene, a synthon that can easily be converted into fuels and several polymers.

The large-scale establishment of cheap chemical processes has allowed the transformation of isobutene into diverse fuels: gasoline, kerosene, diesel and ETBE. These products have been used for decades because of their high energy density and the ease with which they can be handled and stored. These products are fungible in the petrochemical industry of today, unlike ethanol, which would require a complete change in infrastructure if its use were to be generalized. This production process is a biological analog of the Fischer–Tropsch process but does not require a high-temperature step and is therefore better as regards both energy and environment (Global Bioenergies, 2010)[6].

Other futuristic biometabolites. Archaea produce methane. Ethylene is produced at a low level by plants and some microorganisms (*Pseudomonas* sp.). It is therefore not more far-fetched than some of the directions proposed previously to consider improving, via SB, the synthesis of alkanes and alkenes by living organisms. It will be important to invest in the analysis and reconstruction of metabolic pathways in the very near future.

1.3.2 Economic Potential

Although large volumes of bioethanol have been produced in the past years and further increases have been projected for the future, its long-term feasibility has been questioned because of its low energy density, its miscibility with water, its corrosive properties, and necessary engine modifications. Other biofuels may have a promising advantage over bioethanol, such as biodiesel and butanol, both of which can be handled in the existing petrol infrastructure and transportation vehicles. Biodiesel has been considered as the primary renewable alternative to diesel, which is globally in the greatest demand and has a growth rate three times

5) http://www.global-bioenergies.com/index.php?lang=en
6) www.global-bioenergies.com

that of the gasoline market (Steen *et al.*, 2010). The US Defense Advanced Research Project agency spent $2 million in 2009 for **bio-butanol** production research (DARPA, 2010). Glycerine is a major byproduct of biodiesel production. The rapid increase in biodiesel production capacity in Europe has been associated with a correspondingly rapid increase in the amount of (crude) glycerine. Since 2004, glycerine production has exceeded consumption, and the mismatch is increasing. As a result of increasing biodiesel production, and a lack of viable market outlets for glycerine, the price of crude glycerine plummeted in the ten years from the mid1990s. In 2006, the price stood at less than 100€/t (5.8€/GJ), which is only slightly higher than its energy value. At this price level it is too costly for small- and medium-scale biodiesel producers to refine their own crude glycerine. These producers are therefore desperately seeking new glycerine applications. Thus, the leading stakeholders in the European biodiesel sector are urgently seeking to identify new (crude) glycerine applications. The supermethanol project, funded by EC-FP7, aims to identify a viable alternative application by investigating whether (crude) glycerine can be reformed in supercritical water for syngas production and conversion into methanol.

Third-generation biofuels include alcohols like **bio-propanol** or **bio-butanol**, which due to current lack of production experience are usually not considered to be relevant as fuels on the market before 2050 (OECD/IEA, 2008). Increased investment, however, could accelerate their development. The same feedstocks as for first-generation ethanol can be used, albeit using more sophisticated technology. Butamax (Dupont+BP) anticipates that its first commercial plant will be operational by 2013 because its fuel already conforme to European Union standards.

Propanol can be derived from chemical processing such as dehydration followed by hydrogenation.

As a transport fuel, butanol has properties closer to gasoline than bioethanol (UNEP, 2009).

1.3.3
Environmental Impact

The environmental impact of biodiesel has been studied by Hill *et al.* (2006), revealing major advantages over corn-derived bioethanol regarding environmental benefits. By one estimate, biodiesel can reduce GHGs emission by 41% compared with diesel, reduces several major air pollutants, and only minimally impacts human and environmental health through nitrogen, phosphorus and pesticide release. One caveat needs to be mentioned here: biomass often presupposed fertilization using nitrogen, and one underestimated GHG pollutant derived from inappropriate use of nitrogen, namely nitrous oxide, has an enormous and rapidly increasing greenhouse effect. The same estimate points out that corn-based bioethanol provides smaller benefits through a 12% reduction in GHGs. The positive effects of biodiesel on air pollution have also been shown in many studies (Hill *et al.*, 2006). For the non-ethanol biofuels derived from non-food feedstocks (such as lignocellulose), the impacts are comparable to those of bioethanol with regard to greenhouse gas emissions, land use, water consumption, air pollution, and biodiversity. Although biodiesel can be used in conventional engines without major modification, the chemical composition of biodiesel differs distinctly from conventional diesel. Current production of biodiesel comprises a mixture of methyl esters, which is made from the "transesterification" of plant oils such as from rapeseed, soybean or palm oil, using methanol, usually derived from fossil fuels. The process reduces the viscosity of the oil, improves its consistency and miscibility with diesel, and also improves other properties such as its viscosity when cold (AMEC, 2007). As the chemical composition of the oils from each plant species are slightly different, the properties of the final product also differ, and blends of the various oils may be needed to produce a standard. A better fuel could be produced using bioethanol for the transesterification. This would yield ethyl esters that are less viscous, particularly when cold (Kleinov *et al.*, 2007). Owing to minor levels of contaminants and the performance of different processes, there is still some variability in quality. This is more pronounced when waste oils and fats are used as the raw material. Here, the fatty acid content needs to be completely neutralized and either removed or converted to ensure complete reaction and a clean product. Biodiesel is currently limited to 5% in diesel in Europe due to concerns over engine warranties, materials, cold weather performance, and compatibility.

Another problem that needs consideration is the way the initial ligno-cellulosic material meant to be transformed into biofuels is made available: what are the techniques permitting its transport or its cracking. As long as agriculturally derived biomass is used, the situation parallels that of bio-ethanol.

1.3.4 Foreseeable Social and Ethical Aspects

Non-ethanol biofuels are a mixture of very different compounds that can impact society very differently. An evaluation of the impact of such biofuels must first deal with the certainty with which biofuel will predominate in the future. Lacking this knowledge, a case by case estimation of the societal impact can be made:

Bioalcohols. Biobutanol can be used directly as a car fuel instead of gasoline. Its use would therefore imply important savings in terms of delivery and for the automobile industry in particular (no engine modifications required). Additionally, it is less corrosive than ethanol. This as well as other non-ethanol bioalcohols, if produced at a global consumption scale, would imply important savings and recycling of resources and technical material associated with petrol.

Biodiesel. Past experience with the vast use of biodiesel, particularly in city buses and trucks, and the ease of use in standard engines (as in the case of biobutanol) points to a low societal impact of a future large-scale biofuels industry based on biodiesel. The expected substitutions, however, are not gasoline by diesel, but gasoline by bioethanol and diesel by biodiesel. In all cases, the societal impact would be low.

Biogases (biogas and syngas). Gas biofuels can be produced from organic wastes, even locally. Their general use would imply deep societal changes if a local rather than a ready to use delivered fuel economy was set in place.

Solid biomass. As in biogases, solid biomass management might have an important impact because of local production. Additionally, coupling waste processing, such as home garbage or agricultural waste, with energetic production would amount to a turnover in current energy policy and imply obvious societal changes. These include the creation and destruction of jobs linked to waste management, and the transformation, design and trade of the required equipment, and so on.

1.3.4.1 Impact on Social Interaction

The successful application of SB to non-ethanol production (Khalil and Collins, 2010) might help create new businesses including substrate management (organic and agricultural waste, sawdust, etc), promote the development and trade of the required techniques and equipment, and establish a link between both areas. Another issue is the conversion of infrastructures such as pipelines; the specifications of such pipelines will depend on the type of biofuel being transported due to major differences in their corrosive properties.

Ethical issues. Compared to bioethanol, there are no major differences in terms of ethics between non-ethanol and ethanol biofuels. In some cases, the ease of local production for local consumption has to be taken into account (see next point)

Justice of distribution. Two very different scenarios arise: one with massive production of bioalcohols or biogas and ulterior transportation, and another in which large factories are complemented with local production for local use. This would particularly benefit exotic (in the sense of distant) farmers, who might use their agricultural waste as fuel by burning it, use their animal feces for biogas production or produce biodiesel from vegetable oil. The success of direct use of edible oil in old diesel cars (i.e., Mercedes) suggests that taking advantage of the energetic potential of common waste as biofuels does not necessarily require complex reactions. Most of these home-made biofuels might well challenge petrol in terms of dependency. Although these local strategies might be difficult to spread, they would certainly arise in developing countries. Such strategies imply a shift from less sustainable biofuels based on grain, such as corn, towards sustainable feedstock and production practices (Solomon, 2010).

1.4 Algae-based Fuels

1.4.1 Introduction

Algae fuel, also called oilgae, is a biofuel from algae and addressed as a **third-generation biofuel** (UNEP, 2009). **Microalgae contain lipids and fatty acids** as membrane components, storage products, metabolites and sources of energy. **Algal oils possess characteristics similar to those of animal and vegetable oils** and can thus be considered as potential substitutes for the products of fossil oil. Algae are feedstocks from aquatic cultivation for production of triglycerides (from algal oil) to produce biodiesel. The processing technology is basically the same as for biodiesel from second-generation feedstocks. While many microalgae species are capable of producing **high amounts of lipids** (lipid contents exceed those of most terrestrial plants), **higher lipid concentrations** can be obtained in a **nitrogen-limited environment**.

The concept of using algae to make fuel was first discussed more than 50 years ago, but a concerted effort began with the oil crisis in the 1970s (Hu *et al.*, 2008). The US Department of Energy (DOE) from 1978 to 1996 devoted $25 million to algal fuels research in its aquatic species program at the National Renewable Energy Lab (NREL) in Golden, Colorado. The program yielded important advances that set the stage for algal biofuel research today (Waltz, 2009). The first genetic transformation of microalgae came in 1994, and scientists a few years later successfully isolated and characterized the first algal genes that express enzymes thought to enhance oil production. From 1990 to 2000, the Japanese government funded algae research through an initiative at the Research Institute of Innovative Technology for the Earth (Kyoto). The program focused on carbon dioxide fixation and improving algal growth with concentrated mirrors that collect light. These approaches yielded some successes, and many are still the focus of scientists today, but none have proven to be economical on a large scale. The DOE program closed in 1996, in part because algal systems could not compete with the cheap crude oil of the late 1990s. Since the mid1990s, however, the tools for genetic engineering have improved, and scientists are increasingly applying them to algae with fuel applications in mind. Much of the work is focused on identifying the genes involved in lipid synthesis and how those genes are regulated. The idea is to manipulate those genes so that the organisms' metabolic pathways are tricked into producing storage lipids, even when the algae are not under stressful conditions (which is when they start overproducing oil and lipids; Waltz, 2009).

Microalgae can produce high yields of oil that can be refined into transport fuels such as diesel and jet fuel. The advantages of microalgae biofuels over conventional agricultural biofuels are that they can (Spolaore *et al.*, 2006; Waltz, 2009; Carbon Trust, 2010):

- Produce higher yields of oil per hectare of land;
- Do not require arable land or freshwater to grow and thus do not compete with food crops;
- Produce a higher-quality fuel product;
- Can produce non-fuel high-value products (e.g., biopolymers, proteins, animal feed).

If sustainable and profitable processes can be developed, the potential benefits of these technologies for the common good appear compelling and include the production on nonarable land of biodiesel, methane, butanol, ethanol, aviation fuel, and hydrogen using waste or saline water as well as CO_2 from industrial or atmospheric sources (Beer *et al.*, 2009).

Many microalgae are promising for the production of an enormous variety of compounds, including biofuels (see Figure 1.12). To cultivate these algae and their products, monocultures have to be maintained, typically requiring enclosed photobioreactors. A photobioreactor can be described as an enclosed, illuminated culture vessel designed for controlled biomass production of phototrophic liquid cell suspension cultures (Tredici, 1999). Photobioreactors, despite their costs, have several major advantages over open pond systems.

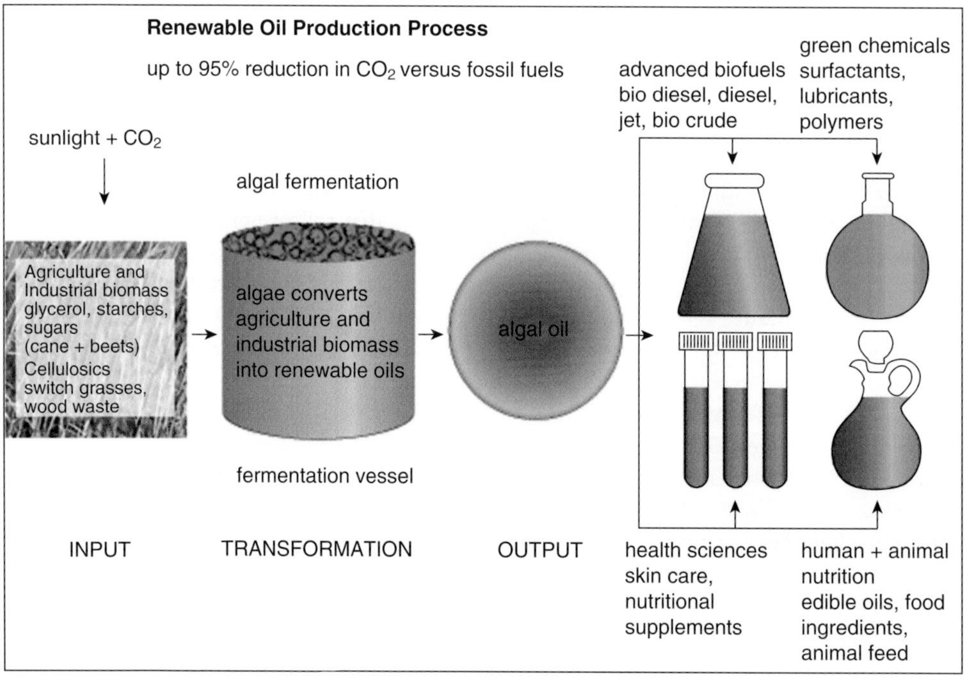

Figure 1.12 Schematic description of the use of microalgae for the production of biomaterials including biofuels. (Source: Solazyme, 2010).

Table 1.8 Productivity of algal strains reported in some outdoor photobioreactors (Ugwu, Aoyagi, and Uchiyama, 2008).

Photobioreactors	Photosynthetic strain	Daily productivity (g/l)
Airlift tubular	*Porphyridium cruentum*	1.5
Airlift tubular	*Phaeodactylum tricornutum*	1.2–1.9
Inclined tubular	*Chlorella sorokiniana*	1.47
Undular row tubular	*Arthrospira platensis*	2.7
Outdoor helical tubular	*Phaeodactylum tricornutum*	1.4
Parallel tubular (AGM)	*Haematococcus pluvialis*	0.05
Bubble column	*Haematococcus pluvialis*	0.06
Flat plate	*Nannochloropsis* sp.	0.27

They can:

- Prevent or minimize contamination, permitting axenic algal cultivation consisting of only one species of microalgae;
- Offer better control over biocultural conditions (pH, light, carbon dioxide, temperature);
- Prevent water evaporation;
- Lower carbon dioxide losses due to gassing;
- Permit higher cell concentrations.

However, certain requirements of photobioreactors–cooling, mixing, control of oxygen accumulation and biofouling–make these systems more expensive to build and operate than ponds. New, cheaper, innovative systems are being designed, and waste streams are used to make the production of microalgae commercially attractive. (See Table 1.8 for details on the productivity of different photobioreactors). Although algae are commercially cultivated in developed countries, most of these regions are characterized by seasonal variations in temperatures and solar light energy over the year. Tropical developing countries might be better potential cultivation sites for the commercial production of algal products (Ugwu, Aoyagi, and Uchiyama, 2008).

1.4.2 Economic Potential

Algae have long been thought to be a rich and ubiquitous source of renewable fuel, but thus far have failed to be economically competitive with other more conventional sources of energy. The question is, what are the major limitations for economic competitiveness and how can they be overcome?

In the United Kingdom the Carbon Trust (2008) launched the Algae Biofuels Challenge in 2008, a research and development investment strategy that could help make low-cost algae biofuels a commercial reality by 2020. The total program cost is estimated to be within £20–30 m. The two phases of this initiative are: (i) address fundamental research and development challenges and (ii) a pilot-scale demonstration of algal oil production.

Several studies have demonstrated that the energy yield of microalgae is significantly higher than that of energy crops (see Table 1.9). Depending on the achievable photosynthetic conversion effiency, microalgae systems could reach even higher oil biomass yields, as seen in Table 1.10.

A detailed economic analysis of microalgae production was carried out by Stephens *et al.* (2010). Two models were used, the base case and the projected case. The base case is intended to represent an emerging scenario from the industry and involves the following assumptions: (i) production of microalgal biomass using 500 ha of microalgal production systems, (ii) the extraction of oil, (iii) the co-production and extraction of a high-value product (HVP; e.g., β-carotene at 0.1%

Table 1.9 Potential oil yields of different crops per year (Waltz, 2009).

Crop	Annual yield (liters of oil/ha)
Soybeans	402
Sunflower	804
Canola	1600
Jatropha	2002
Palm oil	5996
Microalgae[a)]	56124

a) With future technology.

Table 1.10 Practical and theoretical yield maxima for microalgal biomass and oil production (Waltz, 2009).

Photosynthetic conversion efficiency	Annual biomass energy	Biomass energy	Oil	Daily biomass production	Annual biomass yield	Annual oil yield	Annual residual biomass
(%)	(GJ/ha)	(MJ/kg)	(%)	(g/m^2)	(T/ha)	(l/ha)	(T/ha)
2.1	1677	22.98	25	20	73	19837	55
6.5	5220	22.98	25	62.2	227	61400	170
6.5	5220	27.95	50	51.2	187	100943	93
8	6424	22.98	25	76.7	280	75570	210
8	6424	27.95	50	63	230	124237	115
10	8030	22.98	25	95.9	350	94462	262
10	8030	27.95	50	78.6	287	155297	143

of biomass, $600/kg), and (iv) the sale of the remaining biomass as feedstock (e.g., soymeal or fishmeal substitute). In contrast, the projected case is intended to represent the microalgal biofuel industry at maturity and no longer incorporates the co-production of HVPs. The internal rate of return (IRR)[7] has to be above 15% in order for the investment to be profitable, which can be met in the base case but can only be achieved in the projected case (fuel production only) when the oil price exceeds $100 per barrel. Economic feasibility for the microalgae industry seems to be restricted to large industry because approximately 200 ha are needed to be profitable (not including government taxes and other regulatory mechanisms to favor new and renewable energies; see Figure 1.13). Considerable synergies also exist between microalgae biofuel production and a wide range of other industries, including human and animal food production, veterinary applications, agrochemicals, seed suppliers, biotech, water treatment, coal seam gas, material supplies and engineering, fuel refiners and distributors, bio-polymers, pharmaceutical and cosmetic industries, as well as coal-fired power stations (CO_2 capture) and transport industries, such as aviation. Sound opportunities therefore exist for the development of a rapidly expanding sustainable industry base (see Table 1.11) whose productivity is independent of soil fertility and less dependent on water purity. Thus, these technologies can conceivably be scaled to supply a substantial fraction of oil demand without increasing pressure on water resources while potentially contributing to food production.

For the algae-fuel companies the goal is a fuel production at $50–60 per barrel within three to five years. The Defense Advanced Research Projects Agency, which is developing new jet fuel technology for use by the military, is targeting $1 per gallon algal oil, or $1 per gallon finished cost of jet fuel at a capacity of 189 million liters per

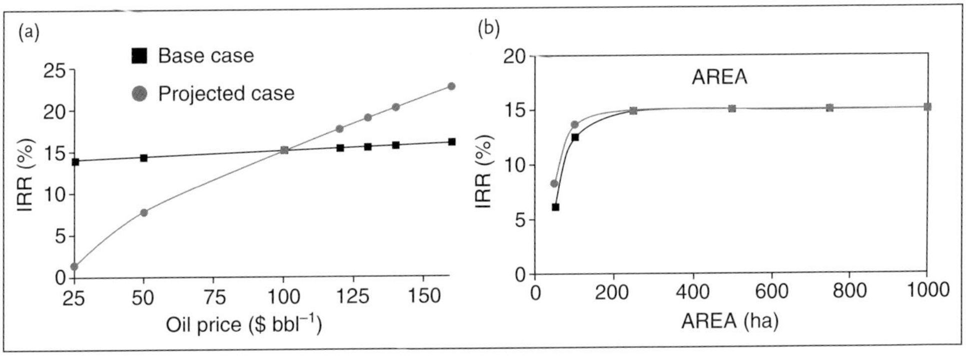

Figure 1.13 Results of a calculation estimating the economic feasibility of microalgae systems. Two cases were calculated, a base case and a projected case. (a) Internal rate of return depending on current oil price. (b) Internal rate of return depending on size of algae factory. The internal rate of return needs to be higher than 15% to be profitable (Waltz, 2009).

7) The IRR of an investment is the interest rate at which the costs of the investment lead to the benefits of the investment.

Table 1.11 Overview of companies that work on biofuel production from algae, applying different genetic techniques (source: Waltz, 2009).

Company	Technology investment	Investors
Aurora Biofuels	Overexpressing genes, such as carboxylic acetylcoenzyme A, to improve triglyceride synthesis	US$ 25 million; Gabriel Venture Partners, Noventi, Oak Investment Partners
Algenol	Metabolically enhancing cyanobacteria to directly synthesize ethanol; expressing enzymes pyruvate decarboxylase and alcohol dehydrogenase II	US$ 70 million; firm's founders; license with Biofields (Mexico) for undisclosed fee and royalties
Sapphire Energy	Producing a "crude-like oil" that can be refined into gasoline or jet fuel	US$ 100 million in 2008; ARCH Venture Partners, Wellcome Trust, Cascades Investments Gates
Solarvest BioEnergy	Hydrogen production in a single cycle from algae; plastid engineering through introduction of promoters that target proteins involved in photosynthesis, hydrogenases and proteins involved in their regulation	Publicly traded on the TS X Venture Exchange
Solazyme	Use of antisense and RNA interference to regulate light-harvesting genes, chlorophyll biosynthesis genes and signaling, together with synthetic genes containing unnatural codons to maximize triglyceride production through fermentation. Business plan is to target nutraceuticals market, then biofuels	US$ 70 million; Roda Group, Harris & Harris Group, Lightspeed Venture Partners, Chevron Technology Ventures
Synthetic Genomics	Modifying genes to create new secretion pathways through the outer membranes of algae so they expel the oil, making harvesting easier. Patent describes genome assembly technology in which cassettes of algal genes involved in triglyceride productions are cloned and rebooted into appropriate recipient prokaryotic cells	Valued at US$ 300 million Investors include founders, Exxon Mobile, BP, Biotechonomy, and others

year (mlpy) or 50 million gallons per year (mgpy) by the end of 2011. However, the current price per gallon for algae fuel is significantly higher, and significant hurdles need to be overcome before algae-fuel becomes economically viable.

Solazyme is the most mature company; it has developed products that met several United States and European standards for biodiesel, renewable diesel and jet fuel, and they envision commercialization within fiveyears. Their fuels have been tested in unmodified engines and meet US Department of Defense specifications. For companies using phototrophic algae, commercialization is not expected for another 10 years.

Two main biological factors govern the cost of algal technologies. The first is the quick selection of the best algal strains among the 40 000 species of microalgae known. The second is engineering the metabolic pathways that control oil production in algae organisms. The genetic modifications can be used to increase the algae's photosynthetic efficiency, growth rate and oil content, and to identify and control the biochemical triggers that cause the organism to accumulate oil. The potential of genetic engineering technology has led some industry experts to conclude that it is a certainty. Several companies, including Algenol, Solazyme, Seambiotic, Sapphire Energy, and Hawaii-based Kuehnle AgroSystems, are working on strains of genetically improved algae.

1.4.3
Environmental Impact

Algae-based biofuel has a high potential to replace a significant proportion of petroleum used in transportation, which will reduce comsumation on tremendous amount of fossil carbon every year globally. Some reports conclude that microalgae are the only source of renewable biodiesel that can meet global demand for transportation fuel (Chisti, 2008).

Initial forecasts suggest that algae-based biofuels could supplement 70 billion liters of fossil-derived fuels used worldwide annually in transportation by 2030. This would equate to an annual carbon saving of over 160 million tonnes of CO_2 globally and a market value of over €17 (£15) billion. If successful, algae could deliver six to 10 times more energy per hectare than conventional cropland biofuels, while reducing carbon emissions by up to 80% relative to fossil fuels. Also, unlike traditional biofuels, algae can be grown on non-arable land using seawater or wastewater. Many species of algae thrive in seawater, water from saline aquifers, or even wastewater from treatment plants. It is believed that the production of algae-based biofuel will not compete for the scarce arable land and can remove nutrients and contaminants from waterways. Therefore, using algae as a biofuel feedstock avoids many of the negative environmental and ecological impacts associated with first-generation biofuels (Carbon Trust, 2010). Algal biofuels offer environmental benefits by reducing anthropogenic pollutant release to the environment and by requiring fewer water subsidies (Smith *et al.*, 2009). Moreover, the algae production unit can potentially be **coupled to an industry that emits CO_2**, thus also **limiting CO_2 emission**. At full scale, algae ponds are estimated to be capable of consuming approximately 80 tonnes of CO_2 per acre per year. For heavy carbon emitters such as steel and coal-fired power plants, cement factories and manufacturing facilities, PetroAlgae offers a cost-effective process to meet increasingly stringent global carbon standards. Also, with micro-crops able to absorb approximately twice their weight in CO_2, PetroAlgae sees carbon management as another potentially profitable enterprise for the future.

1.4.4
Foreseeable Social and Ethical Aspects

Although the assumption that all "land-grown" biofuels adversely affect the supply of food and other crop products is not always true (particularly in case of lignocellulosic bioethanol), it is a fact that aquatic biofuel cultures should not compete with crops. Note, however, that large artificial or semi-artificial aquatic extensions used for culturing algae might indeed compete with another type of human food: fish and seafood from aquaculture. Thus, the societal impact of a hypothetically large algae-based biofuel industry will depend on the interaction between the industry with aquaculture and agriculture (see next issue), and will also be influenced by its ecological effects, such as transformation of wetlands into biofuel factories (see Figure 1.14).

1.4.4.1 Could the Application Change Social Interactions?

Culturing algae is very different from agriculture. Should SB approaches boost the already high lipid production rates of green algae such as to promote their use as biofuel producers, a series of social interaction changes might occur. One scenario

Figure 1.14 High light requirements might require vast surfaces to be transformed for algal culturing (Source: Dasolar Energy, 2010).[12]

12) See: http://www.dasolar.com/alternative-energy/biofuels

envisages that farmers on suitable lands (mainly coastal areas but also those living near wetlands and lakes) will to shift to green algae culturing. This would have important social effects in terms of the industrial structure of the respective societies. Moreover, the local- versus monopoly-based economy represents two non-exclusive future possibilities, as in the case of other biofuels. In the case of green algae, however, local production would be easier than local transformation of algae to biodiesel.

Ethical issues. Optimization of the surface to volume ratio in order to allow proper illumination of algal cultures might require vast surfaces for growing the algae. Note that the conflict between algal culturing, agriculture and aquaculture is not merely a technological issue but should be regulated with a suitable legal framework.

Justice of distribution. Local production is indeed possible, but local transformation might be poor in developing countries. Another expectation is that the technical requirements to efficiently illuminate algal cultures will not be met in developing countries, implying a lower productivity even under tropical climates. Although difficult to forecast, a reasonable scenario is that both local and specialized production coexist, although with a predominant role of the latter in terms of production and, particularly, processing.

1.5 Hydrogen Production

1.5.1 Introduction

The need to seek alternative sources of energy has become a question of major relevance and serious social concern because fossil fuels have limited reserves and adverse environmental effects. Thus, several biofuels such as biodiesel or ethanol have been proposed as suitable candidates to partially substitute oil, although their production potentials are currently limited. Massive efforts are currently underway to enlarge the spectrum of efficiently usable biomass from cellulose. Nonetheless, other alternatives have to be found in the medium and long term. Currently, hydrogen is considered to be a most promising candidate as a future energy source due to its marked reduction to CO_2 emission and extraordinary energy density (142 MJ/kg for H_2 vs 42 MJ/kg for oil).

At the present time, steam reforming of natural gas is the best established system to produce hydrogen on the industrial scale; however, it is based on non-renewable energy sources and generates substantial sulfur and CO_2 emissions. In fact, most of the H_2-producing processes based on fossil fuels release approximately twice as many moles of CO_2 for each mole of hydrogen. In order to produce hydrogen from renewable and clean sources of energy, several alternatives are

Table 1.12 Overview of different hydrogen production processes, including photosynthetic biohydrogen.

Hydrogen production process	Basic technical principle	Positive aspect	Challenge
Coal gasification	Coal is exposed to hot steam and breaks down into a mixture of gases including hydrogen	Technology is available now	Needs high temperature (>700 °C) and energy, CO_2 pollution
Thermochemical	At high temperature various chemical reactions can split hydrogen and oxygen	Heat could be generated by concentrating sunlight	Need to find the most efficient among dozens of candidate reactions
Photo-electrochemical	A single electrode absorbs solar energy and splits water by electrolysis	More efficient than electrolysis of water using power from solar cells	Need to find materials that work well, but do not corrode
Biomass-fermentation	Some bacteria can metabolize cellulose and produce hydrogen	Cellulose is available in large quantities in crop waste	Bacterial metabolism must be re-engineered to make process more efficient
Biohydrogen farming	Cultures of algae/cyanobacteria produce hydrogen using sunlight	The most environmental friendly form of hydrogen production	Photosynthesis must be re-engineered to produce more hydrogen rather than sugars

under study. Among them, biological hydrogen production is considered a promising alternative provided that several limitations are overcome. Hydrogen is being promoted as a potential transport fuel, fulfilling the energy needs of buildings and portable electronics (instead of batteries). Free hydrogen does not occur naturally in large quantities and must thus be generated by various processes (Table 1.12). In contrast to carbon or carbohydrates, hydrogen is only an energy carrier (like electricity), but not a primary energy source. Photosynthetic biofuels entail the direct application of photosynthesis to generate biofuels. In this process, a single organism acts both as catalyst and processor, synthesizing and secreting ready to use fuels. Diverting the natural flow of photosynthesis in autotrophic organisms can generate hydrogen and hydrocarbon gas, instead of the normally produced oxygen. The characteristic of this approach is product generation directly from photosynthesis, and spontaneous product separation from the organism, bypassing the need to harvest and process the respective biomass (Melis, 2007).[8),9)] Modification of photosynthesis in green microalgae may enable the generation of these biofuels as clean, renewable and economically viable commodities (Melis and

8) http://epmb.berkeley.edu/facPage/dispFP.php?I=25
9) This holds also true for gaseous hydrocarbon fuels such as isobutene, as demonstrated, e.g., by the company Global Bioenergies.

Happe, 2002). However, specific biological problems associated with a sustained, high-yield photosynthetic production of these biofuels remain to be addressed.

Currently, the following types of H_2 photoproduction processes are known (for more details see Table 1.12; Tamagnini *et al.*, 2002; Schütz *et al.*, 2004; Ghirardi *et al.*, 2009):

1) Oxygenic photosynthesis:
   - Green algal [FeFe]-hydrogenases;
   - Cyanobacterial [NiFe]-hydrogenases;
   - Cyanobacterial nitrogenases.
2) Non-oxygenic photosynthesis:
   - Bacterial nitrogenases.

Current **objectives of H_2 photoproduction** are (Pulz, 2001; Richmond, 2004; Stripp *et al.*, 2009):

Maximize the solar to chemical conversion efficiency of photosynthesis under mass culture conditions;
Improve the continuity and yield of the green microalgal hydrogen and hydrocarbon production;
Develop advanced tubular photobioreactors for biofuel production;
Enable hydrogen production in the presence of oxygen (engineering O_2-tolerant hydrogenase).

Algae such as *Chlamydomonas reinhardtii* will switch to the photosynthesis pathway under murky conditions, without the full glare of the sun. When sunlight is scant, the arrays of chlorophyll and other pigments will be actively organized into structures termed "antenna complexes" that are remarkably efficient at absorbing sunlight. In an hydrogen farm with plenty of sun, such extensive antenna complexes would not be necessary, but in contrast, would prevent sunlight from reaching the cells located in the center. Scientists are now attempting to engineer algae that contain less chlorophyll. The antenna complexes of normal *C. reinhardtii* cells contain a total of 470 chlorophyll molecules, but they should still be able to photosynthesize if these were stripped down to just 132 chlorophylls – scientists have calculated that this would increase a hydrogen farm's productivity by a factor of four. Unfortunately, that alga does not yet exist. In creating a strain that has these properties, researchers are going through the laborious process of making thousands of mutant *C. reinhardtii*. This involves disrupting their genes by inserting marker sequences of DNA into the cells, which become randomly incorporated into the genome. So far, several promising mutants have been identified (Hemschemeier and Happe, 2005; Tetali, Mitra, and Melis, 2007; Surzycki *et al.*, 2007).

The **ideal algal hydrogen production system** would meet the following criteria:

- It would have no cell wastage – the cells would naturally maintain the same cell density without a net increase in cell mass, and new cells would obtain nutrients through cryptic growth off dead cells.

- The pond depth would be just enough to maximize light adsorption, but no deeper.
- The cells would have a reduced antenna complex size so that they would adsorb only as many photons as they could convert to hydrogen and no more. This would allow the additional photons to pass deeper into the algal solution and be adsorbed by cells further down in the liquid. All the incident photons would thereby be absorbed and converted to hydrogen.
- All electrons passing through photosysthetic system II (PS II) would be used for hydrogen production, with no side reactions.
- The cells would produce hydrogen at the maximum rate at which they could process electrons, without any concerns about oxygen production and inhibition.
- The cells would be contained in a cheap, durable translucent reactor material that fully transmitted all required wavelengths, would have a low hydrogen permeation rate to contain the hydrogen, and would not allow algae cells to attach to the inner surface and block the sunlight.

1.5.2
Economic Potential

The hydrogen economy is defined as an economic system depending on hydrogen-based energy. Hydrogen could be used as an alternative fuel system to hydrocarbons or batteries and therefore has an immense economic potential. The utility of a hydrogen economy depends on a number of issues, including use, availability and costs of fossil fuels, climate change, efficiency of the hydrogen production, and policies for sustainable energy generation. The proposed replacement of the traditional fossil fuel economy by a hydrogen economy would, however, require a massive change in infrastructure, as the pipelines, tanks and motors of these two systems are not compatible (Rifkin, 2002; Garman *et al.*, 2003; Kammen *et al.*, 2003; NAE, 2004). The feasibility of an hydrogen economy depends also on the price of hydrogen production; see Tables 1.13 and 1.14 for details.

1.5.2.1 Cost Comparison with Gasoline for Transport Fuels

In a recent test a Toyota Highlander Hybrid drove 110 km with 1 kg hydrogen under real-world testing conditions.[10] Compared to internal combustion engines in cars, the estimated price (including taxes) of gasoline in Europe is 1.2€/l. Used in rather fuel-efficient cars (100 km per 6 l), the cost for 110 km is 7.92€, which means that at current costs (not taking into account the different costs of cars with fuel cells), hydrogen would have to be cheaper than 7.92€ to be more cost-effective than gasoline in cars (Table 1.15).

The price alone, however, is not the only requirement for a viable hydrogen transportation system to emerge. The US National Academy of Engineering (2004)

10) See http://multivu.prnewswire.com/mnr/toyota/39419

Table 1.13 Characteristics of the three major biological H_2 photoproduction processes.

	Green algae and cyanobacteria (hydrogenase-based)	Cyanobacteria (nitrogenase-based)	Purple bacteria (non-oxygenic, nitrogenase-based)
Light absorption spectra	400–700 nm	400–700 nm	400–600 nm and 800–1010 nm
Photons/H_2 generated	4	15	15
Estimated maximum light conversion efficiency (EMLCE)	10–13%	6%	6%
Electron donor	Water	Water	Organic acids

Table 1.14 Estimated cost of hydrogen per kilogram in a variety of scenarios.[a)]

Source	Hydrogen selling price ($/kg)
Norwegian hydrogen gas station in 2008	6.28
Hydrogen from natural gas (produced via steam reforming at fueling station)	4.00–5.00
Hydrogen from natural gas (produced via steam reforming off-site and delivered by truck)	6.00–8.00
Hydrogen from wind (via electrolysis)	8.00–10.00
Hydrogen from nuclear (via electrolysis)	7.50–9.50
Hydrogen from nuclear (via thermochemical cycles – assuming the technology works on a large scale)	6.50–8.50
Hydrogen from solar (via electrolysis)	10.0–12.00
Hydrogen from solar (via thermochemical cycles – assuming the technology works on a large scale)	7.50–9.50
US DOE future pricing goal	2.00–3.00

a) http://www.reuters.com/article/idUSTRE54A42Z20090511?pageNumber=1&virtualBrandChannel=0; http://www.h2carblog.com/?p=461; http://www1.eere.energy.gov/hydrogenandfuelcells/news_cost_goal.html; http://www.microbemagazine.org/index.php?option=com_content&view=article&id=309:photobiological-hydrogen-productionprospects-and-challenges&catid=132:featured&Itemid=196

summarized four major challenges that would have to be solved for a hydrogen economy to be possible. They concluded that it is necessary:

- To develop and introduce cost-effective, durable, safe, and environmentally desirable fuel cell systems and hydrogen storage systems.
- To develop the infrastructure to provide hydrogen for vehicle users.[11)]

11) See http://www.fuelcells.org/info/charts/h2fuelingstations.pdf for a list of worldwide hydrogen fueling stations.

Table 1.15 Capital costs, operating costs and projected hydrogen selling prices for different algal hydrogen system configurations (e.g., 300 kg/day hydrogen output, $10/m^2 reactor cost; PSA: pressure swing adsorption purification; Amos, 2004).

System design	Capital cost ($)	Annual operating cost ($)	Minimum hydrogen selling price ($/kg)
300 kg/day, $100/m^2, PSA, high-pressure storage	22.2 mio	614 000	439.00
300 kg/day, $10/m^2, PSA, high-pressure storage	5.2 mio	119 000	13.53
600 kg/day, $10/m^2, PSA, high-pressure storage	9.1 mio	214 000	11.96
300 kg/day, $10/m^2, PSA, pipeline delivery	3.2 mio	131 000	5.92
300 kg/day, $10/m^2, high-pressure storage	5.0 mio	115 000	12.93
300 kg/day, $10/m^2, pipeline delivery	2.9 mio	127 000	5.52
300 kg/day, $10/m^2, ponds only, no compression	1.9 mio	101 000	3.68
300 kg/day, $1/m^2, ponds only, no compression	0.2 mio	51 000	0.57
300 kg/day, $1/m^2, PSA, high-pressure storage	3.5 mio	70 000	8.97
300 kg/day, $1/m^2, PSA, pipeline delivery	1.5 mio	81 000	2.83

- To reduce sharply the costs of hydrogen production from renewable energy sources, over a timeframe of decades.
- To capture and store ("sequester") the carbon dioxide byproduct of hydrogen production (from fossil fuels such as coal).

Clearly, cheap hydrogen production from renewable energy sources is one important goal and a necessary requirement, but by far not the only one. Carbon dioxide sequestration will also play a role when using fossil fuel for hydrogen production (see also Section 6.3), at least as a temporary energy source to pave the way for a sustainable hydrogen economy. Other challenges such as more efficient fuel cells, storage systems and a proper infrastructure are also key requirements; they cannot be solved by synthetic biology, but require other engineering fields. The hydrogen economy is a complex scientific, technological and political goal: synthetic biology could contribute, but will not be able to solve all the problems by itself. Interdependencies between different science and engineering fields and political decisions will determine whether the hydrogen economy will take off or not.

On 24 March 2009, Sapporo Breweries Ltd announced that the company would start proof production experiments of biohydrogen using agricultural produce like sugarcane. This is a joint project with the Brazilian oil company Petrobras (Rio de Janeiro) and Ergostech Co. (São Paulo), specializing in research and consulting on renewable energy. This proof experiment to produce hydrogen from cellulose-type biomass is the first venture in the world. In fact, Sapporo Breweries Ltd has formerly succeeded in developing a pilot plant for the production of hydrogen-methane in a two-stage fermentation process utilizing waste bread as raw material. This achievement is based on the company's original technology and know-how as regards fermentation as well as processing plant design which have been gained in brewing beer. They plan to install and operate a pilot plant with 1 m^3 capacity at the experimental laboratory of Ergostech by mid-September 2009. In 2010, they intend to conduct a continuous fermentation experiment, utilizing waste from vegetables and crops. They also plan to install a pre-commercial production plant to enable a proof production experiment in 2013 and later. They intend to spend $2.5 million for the project in order to realize, within 10 years, a production cost of 40 Yen/m^3, which would be equal to and competitive with the cost of crude oil and natural gas.

1.5.3
Environmental Impact

Safety issues: Hydrogen ignites easily upon contact with oxygen (air) and a flame/spark. Hydrogen has a wide flammability range (4–74% in air) and the energy required to ignite hydrogen (0.02 mJ) can be very low. At low concentrations (below 10%), however, the energy required to ignite hydrogen is higher – similar to the energy required to ignite natural gas and gasoline in their respective flammability ranges – making hydrogen realistically more difficult to ignite near the lower flammability limit. Nonethelss, if conditions exist where the hydrogen concentration increases toward the stoichiometric (most easily ignited) mixture of 29% hydrogen (in air), the ignition energy drops to about one-fifteenth of that required to ignite natural gas (or one-tenth for gasoline; see Table 1.16; Hydrogen Association, 2010).

Hydrogen is lighter than air and diffuses rapidly. This rapid diffusivity means that, when released, hydrogen dilutes quickly into a non-flammable concentration. Hydrogen rises two times faster than helium and six times faster than natural gas. Therefore, unless an enclosed room contains the rising gas, the laws of physics prevent hydrogen from lingering near a leak. Care has to be taken in enclosed spaces such as buildings, tunnels, underground parking lots and so on (Hydrogen Association, 2010; see Figure 1.15).

The odorless, colorless and tasteless of hydrogen gas make it difficult to detect any possible leak. Therefore, hydrogen sensors need to be used to detect possible leaks. By comparison, natural gas is also odorless, colorless and tasteless, but

Table 1.16 Relevant safety data for three major fuels (source: Hydrogen Association, 2010).

	Hydrogen	Gasoline vapor	Natural gas
Flammability limits (in air; %)	4–74	1.4–7.6	5.3–15.0
Explosion limits (in air; %)	18.3–59.0	1.1–3.3	5.7–14.0
Ignition energy (mJ)	0.02	0.20	0.29
Flame temperature in air (°C)	2045	2197	1875
Stoichiometric mixture (most easily ignited in air; %)	29	2	9

Figure 1.15 The famous Hindenburg disaster in 1937. Hydrogen explodes–the main reason why today's zeppelins are filled with helium. (Source: Shere, 1937).[13]

industry adds a sulfur-containing odorant, called mercaptan, to make it detectable by people. Currently, all known odorants contaminate fuel cells and create complications for food applications. Researchers are investigating other methods that might be used for hydrogen detection, among them tracers and advanced sensors.

Asphyxiation: With the exception of oxygen, any gas can cause asphyxiation. In most scenarios, hydrogen's buoyancy and diffusivity make it unlikely to be confined and potentially cause asphyxiation (Hydrogen Association, 2010).

Toxicity/poison: Hydrogen is non-toxic and non-poisonous. It will not contaminate groundwater (it is a gas under normal atmospheric conditions), nor will a release of hydrogen contribute to atmospheric pollution. Hydrogen does not create "fumes" and is not a greenhouse gas (Hydrogen Association, 2010).

13) See: http://iconicphotos.wordpress.com/2009/07/25/hindenberg-disaster/

1.5.3.1 Environmental Concerns

Hydrogen can be used as an additive in internal combustion engines (ICE); an ICE running on hydrogen may produce nitrous oxides and other pollutants, for example, nitric acid (HNO_3) and hydrogen cyanide gas (HCN). As a trasportation fuel, however, hydrogen is mainly used in fuel cells, not internal combustion engines, thus avoiding the burning pf hydrogen in the presence of nitrogen.

Concerns have also been raised over possible problems related to hydrogen gas leakage and subsequent effects on the atmosphere. Molecular hydrogen leaks slowly from most containment vessels and it has been hypothesized that if significant amounts of hydrogen gas (H_2) escape, hydrogen gas may, because of ultraviolet radiation, form free radicals (H^+) in the stratosphere (Schultz *et al.*, 2003; Tromp *et al.*, 2003). These free radicals would then be able to act as catalysts for ozone depletion. A large enough increase in stratospheric hydrogen from leaked H_2 could exacerbate the depletion process. However, the estimations behind those proposed effects of these leakage problems may not be correct. Tromp *et al.* (2003) note that the amount of hydrogen that leaks today is much lower (by a factor of 10 to 100) than the estimated 10–20% figure conjectured by some researchers. A more realistic value is only about 0.1%, which is less than the natural gas leak rate of 0.7%. Tromp *et al.* (2003) conclude that the effect of hydrogen leakages on the atmosphere and the ozone layer (which we hope, based on current trends, will be largely repaired once the hydrogen economy is in full swing by 2050) will be negligible.

A positive effect of the hydrogen economy and its use in transport (cars) will be the near elimination of controllable urban air pollution by the end of the century (HTAC, 2010). Eliminating current vehicle exhaust will save thousands of lives and would hardly affect tropospheric water vapor concentrations (Jacobson, Colella, and Golden, 2005). Some researchers, however, note that the widespread use of fuel cell cars will reduce urban pollution but will create effects on the microclimate due to the increased water vapor that fuel cells emit (Pielke *et al.*, 2005). Pielke *et al.* conclude that "In the case of hydrogen cars, the cure may indeed be better than the disease, but we should make sure before taking our medicine."

1.5.4 Foreseeable Social and Ethical Aspects

Hydrogen is perhaps one of the alternative biofuels that might have a deeper societal impact. There are many ways to produce hydrogen, and hydrogen is often considered an "energetic carrier" because it can be produced from water (with electricity) and then distributed for energetic purposes. This is one of the bases of the "economy of hydrogen" theory by Jeremy Rifkin. If, however, we consider only biomass-obtained hydrogen, then a real biofuel instead of an energetic carrier arises. The ease of transportation of this biomass-obtained hydrogen and the ease of *in situ* conversion are important advantages. A main concern is the explosive nature of the gas. Irrespective of its origin, the combustion of hydrogen yields only heat and water. No global warming emissions or toxic pollutants are formed (Zeman and Keith, 2008). Thus, the non-carbonic nature of hydrogen makes this the most environmentally friendly of the biofuels. Accordingly, a real "hydrogen economy" would be a turnover of the current carbon-based fuel economy.

1.5.4.1 Could the Application Change Social Interactions? If Yes, in Which Way?

Important changes in terms of social interactions are expected if SB approaches help achieve a complete development of the hydrogen economy. New jobs and businesses related to the technology and especially to the distribution of hydrogen are expected to be created. It is important to note that security issues due to the explosion risk may also require novel control structures, jobs and even terrorist risk assessment.

Two main **ethical issues** mark the difference between hydrogen and carbon-based biofuels. The former is the absence of environmental impact in terms of greenhouse effect when hydrogen is burned [only applies to biomass hydrogen because non-biomass hydrogen production (i.e., electricity-based) does indirectly produce CO_2]. The latter is the improvement in the justice of distribution of hydrogen as an energetic source (see below).

Justice of distribution. If hydrogen was locally produced, it could be used as a fuel without many technical difficulties. Alternatively, if the production of hydrogen was centralized, it could be relatively easily transported through pipelines. One uncertainty is whether developing countries would be able to afford the construction of such structures and the price of the delivery.

Among the major challenges for the hydrogen economy, the US DOE (2010) lists public acceptance. This is because the hydrogen economy will be a revolutionary change from the world we know today. The necessary steps to foster hydrogen's acceptance as a fuel are: education of the general public, training personnel in the handling and maintenance of hydrogen system components, adoption of codes and standards, and development of certified procedures and training manuals for fuel cells and safety.

1.6 Microbial Fuel Cells and Bio-photovoltaics

1.6.1 Introduction

A microbial fuel cell (MFC) is a device that uses the ability of microorganisms to produce electrical power by converting chemical to electrical energy.

Different MFC technologies have been tested in the laboratory at a fast pace, and power densities have reached over $1\,kW/m^3$ (reactor volume) and $6.9\,W/m^2$ (anode area) under optimal conditions (Logan, 2009; see Figure 1.16). The real

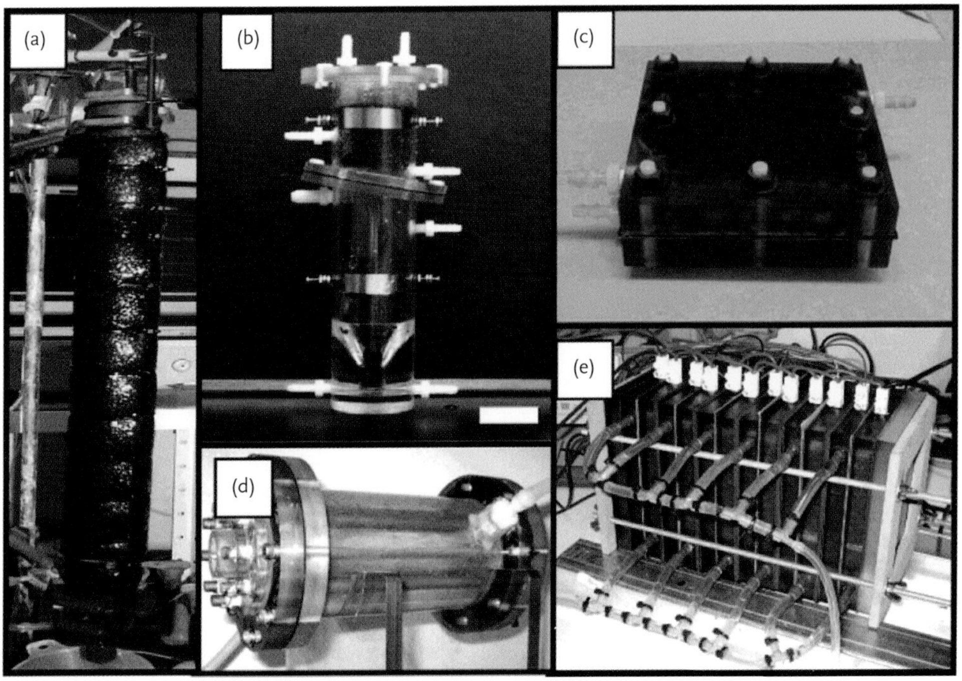

Figure 1.16 MFCs used for continuous operation: (a) upflow, tubular type MFC with inner graphite bed anode and outer cathode; (b) upflow, tubular type MFC with anode below and cathode above; (c) flat plate design; (d) single-chamber system with an inner concentric air cathode surrounded by a chamber containing graphite rods as anode; (e) stacked MFC, in which six separate MFCs are joined in one reactor block (Source: Logan *et al.*, 2006).

challenge, however, is to convert these technologies developed in the laboratory into commercial applications to engineer systems for bioenergy production at larger scales. Recent advances in the global performance of MFCs are the discovery of new types of electrodes and a better knowledge about the role of membranes, separators and nanowires (Reguera *et al.*, 2005; see Figure 1.17). Commercialization of MFCs could start within only a few years (Table 1.17).

MFCs were originally designed to produce electricity (Lovley, 2006), but applications for other purposes also exist. For example, additional voltage added to the potential generated by the bacteria allows the production of methane, hydrogen and hydrogen peroxide. Additionally, membranes can be used in MFCs in such a way that water is desalinated, while electrical power production is kept (Cao *et al.*, 2009).

Since over 100 000 TW of solar energy falls on the Earth every year, solar energy is by far the most abundant primary energy source. The current use of energy on Earth is estimated at 10 TW per year and will double to 20 TW by 2050. The present

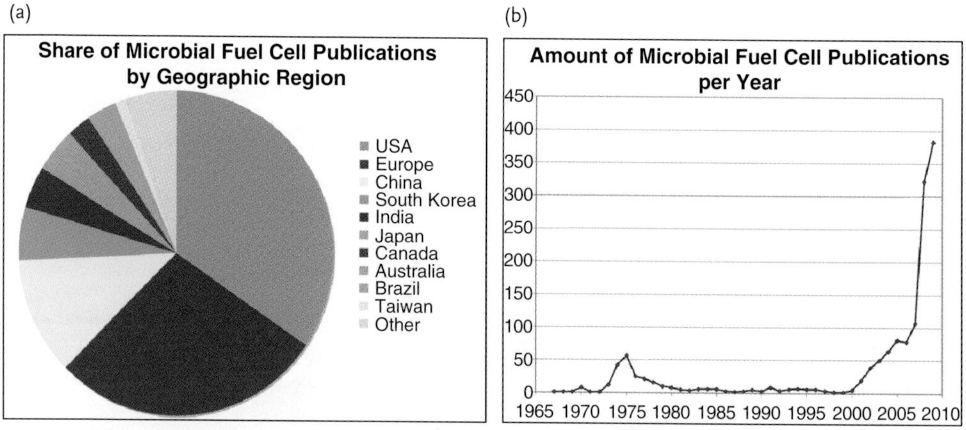

Figure 1.17 (a) In recent years there was a dramatic increase in in the scientific literature about MFCs; (b) the United States published most MFC articles, followed by Europe and China (GoPubMed, 2010).

techniques to convert solar energy directly to electrical energy, namely photovoltaic cells based on semi-conductors, are less effective and expensive. They contain rare elements, making their producibility also depending on uncertain market conditions. Nature possesses highly effective mechanisms for harvesting solar energy. These could be the basis for a new, effective, cheap and renewable photovoltaic system.

In one application, Peter *et al.* (2006) proposed to separate the processes of oxygen evolution and hydrogen production in a semi-biological photovoltaic device using intact photosynthetic cells. Here, protein complexes are intrinsically more stable and have mechanisms for self-repair. The device will be composed of two chambers, or half-cells, with oxygen evolution confined to one chamber and hydrogen production to the other. In addition, the approach can be used to produce a DC electrical current, in a manner analogous to standard silicon-based photovoltaic panels.

Another type of a bio-photovoltaic conversion device is being investigated by Yutaka Amao and Tasuku Komori (Department of Applied Chemistry, Oita University, Japan). It is based on dye-sensitized solar cells (DSSC) using the visible light senitization of chlorine-e_6 (Chl-e_6). The latter is derived from chlorophyll from *Spirulina* adsorbed on a nanocrystalline TiO_2 film.

Developing an alternative technique, the Department of Chemical Engineering and Biotechnology of the University of Cambridge is working on a bio-photovoltaic device which exploits the photosynthetic apparatus of biological material such as cyanobacteria or algae. The idea is, to convert the solar energy into electrical energy and then use this electrical energy to drive a current or create a potential difference to drive a chemical reaction. These few examples show that extensive research is going on in this key field and that there are numerous interesting approaches.

Table 1.17 Potential microbial fuel cells and applications (Davis and Higson, 2007).

Microbial fuel cell	Fuel	Application
In vivo power supply	Blood glucose and oxygen	Pacemaker, glucose sensor, prosthetic valve actuator power supply
Transcutaneous power socket	Blood glucose and oxygen	Variety of low-power electronic devices
Waste remediation	Process residues	Remediation of process wastes with power recovery
Waste remediation	Urine	Use of wastes for power generation in remote areas
Portable power cell	Alcohol	Portable power supply for mobile telephones or other consumer electronics. Instant recharge times, lifetime on the order of one month between replacements. No precious metals required, therefore readily recyclable or disposable
Biosensors	Target molecule	Can act as a specific biosensor (if enzyme-based) or a non-specific one if microbe based. The latter has a potentially indefinite lifetime
Static power generation	Cellulosic materials	Potentially lignocellulosic materials (e.g., corn stalks, wood) could be broken down and used to directly generate power sustainably
Static power generation	Sewage	Sewage-digesting bacteria have been demonstrated to be capable of generating electricity, and the biological oxygen demand of the fuel itself could help maintain a system in an anaerobic state, but the power levels of for a practical system have not yet been demonstrated
Static power generation	Marine sediment	The biofuel cell demonstrated by Tender *et al.* (2002) is almost immediately applicable to provide long-term power to remote marine electronics
Mobile power generation	Organic materials	Start and forget gastrobot[a)] (energy autonomous robots) operations, if suitable food-locating behaviors can be programmed

a) Gastrobot literally means "robot with a stomach": they are machines that power themselves by digesting food.

MFCs can be classified according to their electron transfer scheme and main bacterial species (Bullen *et al.*, 2007; Lovley, 2008):

- Indirect electron transfer through the interaction of reduced metabolic products with the anode;
- Enhanced electron transfer with artificial mediators: *Escherichia coli*, *Pseudomonas*, *Proteus*, and *Bacillus* species;

- Microorganisms that produce their own mediators: *Shewanella oneidensis, Geothrix ferementans* and *Pseudomonas* species (*Pseudomonas aeruginosa*);
- Direct electron transfer to electrodes: *Shewanella putrefaciens, Aeromonas hydrophila*;
- Oxidation of organic matter with electricigens (benthic unattended generator): *Geobacter sulfurreducens, G. metallireducens, G. psychrophilus, Desulfuromonas acetoxidans, Geopsychrobacter electrodiphilus, Rhodoferax ferrireducens.*

1.6.2 Economic Potential

The economic benefits of electricity from MFC can be considered in two sectors: the automotive sector and special applications (fine mechanics, medical technology, telecommunications, IT, etc.). An application in the automotive sector is conceivable. This would create an enormous turnover because this is a global key industry. Special applications in other fields also bear the potential for billion-dollar-scale turnovers because of the diversity of such applications (e.g., a notebook battery working with alcohol which is converted to water and discarded; see, e.g., Debabov, 2008).

1.6.3 Environmental Impact

Due to global environmental concerns and energy insecurity, there is a need to develop cost-effective wastewater treatment processes and sustainable clean energy sources, preferably without the use of fossil fuel. A MFC has a great potential to solve this problem by generating direct electricity during the oxidation of organic matter. MFCs have recently received increased attention as a means to produce "green" energy from organic wastewater or synthetically prepared carbohydrate substrates (Ghangrekar and Shinde, 2008). Over the medium term, MFCs could become an interesting tool to generate electricity from waste (waters; see Figure 1.18). So far, experiments have been carried out with the following substrates (fuels; Ren, Steinberg, and Regan, 2008; Wang, Feng, and Lee, 2008; Pant *et al.*, 2010):

- Acetate;
- Glucose;
- Lignocellulosic biomass;
- Synthetic or chemical wastewater;
- Brewery wastewater;
- Animal wastewater;
- Starch processing wastewater;
- Dye wastewater;
- Landfill leachates;

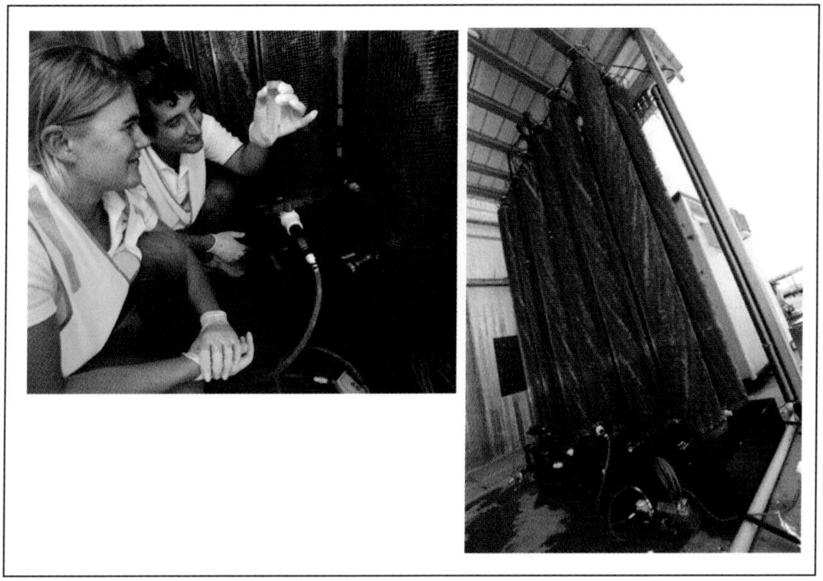

Figure 1.18 MFC pilot plant for brewery wastewater (see: www.microbialfuelcell.org).

- Cellulose and chitin;
- Inorganic and other substrates.

Tested substrates have grown in complexity and content strength (higher organic loading rate). The output of these systems (electric current and electric power) is still some way from large-scale applications. It is believed that more innovations are needed to achieve more technological advancements in terms of materials, costs and substrates will make these systems to be commercially competitive. A promising aspect, however, is that a variety of new substrates (see list above) can serve as substrates under the MFC set up. These include wastewater from molasses-based distilleries rich in organic matter and produced in large volumes, wastewater from a large number of biorefineries, wastewater from the pharmaceutical industry with recalcitrant pollutants, or waste plant biomass (agriculture residue) which is currently being burned. Economically, the integration of MFCs with existing separation, conversion and treatment technologies is probably the best option (Pant *et al.*, 2010). One clear benefit of MFCs is that they can be used as bioremediation tools and electricity generation tools at the same time, working as a direct waste to energy conversion system. Although the obtained power output is still relatively low, the technology is improving rapidly and eventually could be useful to reduce the cost of small sewage and industrial wastewater treatment plants (Ghangrekar and Shinde, 2008). Although waste can fuel MFCs, natural environments can do the same job. For example, MFCs deployed in natural waters can produce enough energy to operate (bio)sensors requiring low power. Note, however, that all these microbial fuel cells produce a maximum cell potential

Figure 1.19 Seal of the United States Navy's benthic unattended generator (BUG).

around 800 mV, which currently limits their use to power electronic devices (Donovan *et al.*, 2008). The **benthic unattended generator** or BUG, developed by the US Naval Research Laboratory's Center for Bio/Molecular Science and Engineering, is a weather buoy that operates in the Potomac River; see Figure 1.19). The buoy is unique in that it is solely powered by a set of microbial fuel cells. These BUGs consist of electrodes imbedded in sediment in the bottom of the river that are electrically connected to electrodes in the overlying water. The buoy monitors air temperature and pressure, relative humidity, water temperature, and performance indicators of the BUGs, and sends data by a radio transmitter (also BUG-powered) to a receiver. Organic matter deposited in many fresh- and saltwater environments constitutes a practically inexhaustible fuel. Furthermore, substantial oxygen present in overlying water constitutes a practically inexhaustible oxidant. BUGs electrochemically react with this fuel and oxygen to generate electrical power that persists indefinitely (as long as the fuel cell lasts). BUGs are being developed to persistently power a wide range of remotely deployed marine instruments. BUG research is partly funded by the Office of Naval Research (ONR) and the Defense Advanced Research Projects Agency (DARPA, 2010). A similar system can also be used in rice paddies. Researchers have developed a microbial fuel cell that allows them to generate up to 330 watts of power per hectare of farmed rice paddies. Taking advantage of a process called rhizodeposition, the MFC can capture some of the energy produced in rice paddies, possibly before it is released in the form of methane. As rice paddies cover well over one million square kilometers of land and are responsible for approximately 10–20% of the world's methane emissions, a method that simultaneously utilizes rice plants to make electricity and reduce methane emissions would have a very positive environmental impact (De Schamphelaire *et al.*, 2008).

The entire system – the microbial fuel cell, sensor, and telemetry systems – may last up to 7.5 years. If needed, this can be doubled or tripled by merely changing the specifications, for example, increasing the numbers of anodes (Shantaram *et al.*, 2005). The advantage to using an MFC as opposed to a normal battery is that it uses a renewable form of energy and would not need to be recharged like a standard battery would. Moreover, they could operate well in mild conditions, 20–40 °C, at a pH of around 7.

1.6.4 Foreseeable Social and Ethical Aspects

One of the most important features of this technology is the possibility of coupling several energetic procedures in the same place. For example, waste treatment, electric power generation, hydrogen production and water desalinization could be obtained at the same time and with the same equipment. This feature makes bio-photovoltaic applications unique, and their massive use would certainly have a deep societal impact by grouping two of the most important factors that hamper human development: energy and waste treatment. As with the other biofuels, new businesses and jobs would arise. Another expected high impact on social interactions would come from the rearrangement of treatment, biofuel and electric power fields into a single technology. Of course, this would only be the case if a significant contribution to these processes was linked to microbial cell fuel-based strategies. Even a relatively reduced spread of the technique, however, would have an impact because of the grouping effect.

Ethical issues. MFC does not have particular ethical issues, except those related to the societal effects of grouping waste treatment and energy policies and those of distribution. Another type of ethical issues could appear when higher order animals would be used to generate electricity, e.g., a dog with MFC that uses the blood glucose and oxygen to power a GPS receiver on its necklace, or cetaceans (dolphins) using a similar system to power electronic sensors searching for enemy submarines.

Justice of distribution. The state of the art of bio-photovoltaic devices consists of laboratory-scale modules that work but produce low voltage/intensities. Assuming a successful development of more productive systems, local production with very simple equipment (see Figure 1.20) would be possible. The outcome of these devices, electrical power, would be ready to use. This aspect, along with the fact that the byproducts of cell fuels can be used as biofuels and the possibility of coupling MFC to water treatment, makes biophotovoltaics one of the most promising technologies in terms of justice of distribution.

1.7 Recommendations for Biofuels

We are convinced that synthetic biology can help to produce state of the art and next-generation biofuels. Current efforts are mainly targeted towards an improved production of **bio-ethanol** from agricultural products, although we see significant problems with this approach because ethanol exhibits certain technical problems

Figure 1.20 Few MFCs are commercially available, but the internet is full of examples of home-made prototypes (source: Makezine, 2006).[14]

(mixes with water, limited use in existing engines). Other non-ethanol biofuels such as **bio-butanol** or **biodiesel** are much better suited to replace petroleum-based gasoline because their chemical properties resemble it much closer. Synthetic biology could help to overcome current impasses in the production of butanol and other non-ethanol fuels, namely poor fermentation yield and toxicitiy to butanol-producing microorganisms. One problem facing most biofuels produced from plant material are limitations on the use of hemi- and lignocellulosic material. Any improvement in that area would definitely increase the economic feasibility of biofuel production. Another important issue will arise should synthetic biology be able to help solve the above-mentioned technical problems: more and more agricultural land will be devoted to plant energy crops instead of food crops. In order to avoid this competition for food, we suggest also using non-food-competing biological resources such as perennial plants grown on degraded lands abandoned for agricultural use, crop residues, sustainably harvested wood and forest residues, double crops and mixed cropping systems, as well as municipal and industrial wastes.

In addition to agriculturally based ethanol, biodiesel and butanol, another option is **algae-based biofuels** and **biohydrogen**. Current concepts foresee a significant advantage of algae-based biofuels over agriculture-based biofuels because of a higher yield per area and the independence from arable land and clean water. Initial calculations predict, however, that future algae poduction systems will be economically feasible only if the price for one barrel of oil remains consistently above \$70 and if the production systems entail an area of at least 200 ha. The capital costs of such large production facilities will probably exlude SMEs and favor "big oil" (or big energy companies). Nonetheless, algae production systems could be a highly promising avenue of future fuel production once major obstacles are resolved dealing with algae genomics, metabolism and harvesting. Although **biohydrogen** has been praised as an extremely promising fuel by many scientists, our assessment is more cautious. Hydrogen is only useful as fuel if large changes in

14) See: http://blog.makezine.com/archive/2006/05/how_to_make_microbial_fue.html

infrastructure take place (distribution and storage system, new fuel cell engines). This points to a more distant future beyond 2050, also termed the hydrogen economy. Although synthetic biology could contribute to improve the yield of hydrogen-producing cyanobacteria, the actual impact of hydrogen in society and economy depends much more on other areas such as infrastructure. Finally, we analyzed the prospects of **microbial fuel cells** (MFCs) as energy converters. Although we see MFCs as extremely promising and a sector in which synthetic biology could make a significant contribution, such fuel cells will most likely be applied in certain niche markets and areas of application, rather than large-scale deployment, due to the limited energy production.

References

1. BIOFUELS

1.1. BIOFUELS IN GENERAL

Biofuels Platform (2010) The situation in the EU: background and objectives, http://www.biofuels-platform.ch/en/infos/eu-background.php (accessed 28 May 2010).

Bringezu, S., Schütz, H., O'Brien, M., Kauppi, L., Howarth, R.W., and McNeely, J. (2009) Towards sustainable production and use of resources: assessing biofuels. United Nations Evrionment Programme.

EC (2003) Biofuels directive 2003/30/EC, http://ec.europa.eu/energy/res/legislation/doc/biofuels/en_final.pdf (accessed 28 May 2010).

EC (2009) Directive 2009/28/EC on the promotion of the use of energy from renewable sources, http://eur-lex.europa.eu/LexUriServ/LexUriServ.do?uri=OJ:L:2009:140:0016:0062:EN:PDF (accessed 28 May 2010).

EC (2010) Market observatory. EU: crude oil imports, http://ec.europa.eu/energy/observatory/oil/import_export_en.htm (accessed 28 May 2010).

Green, C. (2010) Biofuels directive review and progress report: public consultation, http://ec.europa.eu/energy/res/legislation/doc/biofuels/contributions/citizens/green.pdf (accessed 28 May 2010).

IEA (2009) Key World Energy Statistics, http://www.iea.org/textbase/nppdf/free/2009/key_stats_2009.pdf (accessed 28 May 2010).

IEA (2010) Oil market report released, http://omrpublic.iea.org/omrarchive/15jan10full.pdf (accessed 28 May 2010).

SCOPE (2009) Biofuels: environm consequences and interactions with changing land use. SCOPE, Gummersbach. http://cip.cornell.edu/biofuels/ (accessed 28 May 2010).

Tilman, D., Socolow, R., Foley, J.A., Hill, J., Larson, E., Lynd, L., Pacala, S., Reilly, J., Searchinger, T., Somerville, C., and Williams, R. (2009) Beneficial biofuels: the food, energy, and environment trilemma. *Science*, **325**, 270–271.

Thomassen, D.G., Simmons, K., Fernandez-Guiterrez, M., and Lex, M. (2008) EC-US taskforce on biotechnology research: workshop on biotechnology for sustainable bioenergy, http://ec.europa.eu/research/biotechnology/ec-us/docs/us-ec_bioenergy_workshop_proceedings_en.pdf (accessed 28 May 2010).

UNEP (2009) Assessing biofuels, http://www.unep.fr/scp/rpanel/pdf/Assessing_Biofuels_Full_Report.pdf (accessed 28 May 2010).

1.2. ETHANOL

BIO-ERA (2009) U.S. economic impact of advanced biofuels production: perspectives to 2030, http://bio.org/ind/EconomicImpactAdvancedBiofuels.pdf (accessed 28 May 2010).

Domínguez-Escribà, L., and Porcar, M. (2010) Rice straw management: the big waste. *Biofuels Bioprod. Bioref.*, **4**, 154–159.
EEA (2006) How much bioenergy can Europe produce without harming the environment? EEA Report No 7. Office for official publications of the European communities, Luxembourg.
EPA (2010) EPA finalizes regulations for the national renewable fuel standard program for 2010 and beyond, http://www.epa.gov/OMS/renewablefuels/420f10007.pdf (accessed 28 May 2010).
ETC (2008) Commodifying nature's last straw?: extreme genet eng and the post-petroleum sugar econ, http://www.etcgroup.org/en/node/703 (accessed 28 May 2010).
Hatano, K., Kikuchi, S., Nakamura, Y., Sakamoto, H., Takigami, M., and Kojima, Y. (2009) Novel strategy using an adsorbent-column chromatography for effective ethanol production from sugarcane or sugar beet molasses. *Bioresour. Technol.*, **100** (20), 4697–4703.
Melillo, J.M., Reilly, J.M., Kicklighter, D.W., Gurgel, A.C., Cronin, T.W., Paltsev, S., Felzer, B.S., Wang, X., Sokolov, A.P., and Schlosser, C.A. (2009) Indirect emissions from biofuels: how important? *Science*, **326** (5958), 1397–1399.
Rodrigues, L.P., and de Moraes, M.A.F.D. (2007) Estrutura de mercado da indústria de refino de açúcar na região centro-sul do Brasil. *Rev. Econ. Sociol. Rural*, **45**, 93–118.
Sawyer, D. (2008) Climate change, biofuels and eco-social impacts in the Brazilian amazon and Cerrado. *Philos. Trans. R. Soc. B Biol. Sci.*, **363**, 1747–1752.
SCOPE (2009) Biofuels: environm consequences and interactions with changing land use. SCOPE, Gummersbach. http://cip.cornell.edu/biofuels/ (accessed 28 May 2010).
Tilman, D., Socolow, R., Foley, J.A., Hill, J., Larson, E., Lynd, L., Pacala, S., Reilly, J., Searchinger, T., Somerville, C., and Williams, R. (2009) Beneficial biofuels: the food, energy, and environ trilemma. *Science*, **325**, 270–271.
UNEP (2009) Assessing biofuels, http://www.unep.fr/scp/rpanel/pdf/Assessing_Biofuels_Full_Report.pdf (accessed 28 May 2010).
Zah, R. (2007) Umweltauswirkungen von Biotreibstoffen. Presentation at the Energy Science Colloquium Zurich. http://www.esc.ethz.ch/news/colloquia/2007/PresentationZah.pdf (accessed 28 July 2010).
Zeman, F.S., and Keith, D.W. (2008) Carbon neutral hydrocarbons. *Philos. Transact. A Math. Phys. Eng. Sci.*, **366** (1882), 3901–3918.

Interesting link:

A list of cellulolytic bacteria: http://www.wzw.tum.de/mbiotec/cellmo.htm (accessed 28 May 2010)
Cellulosic ethanol. http://en.wikipedia.org/wiki/Cellulosic_ethanol (accessed 28 May 2010)
Food before fuel. http://www.foodbeforefuel.org/facts (accessed 28 May 2010)
Range fuels: biomass to energy. www.rangefuels.com (accessed 28 May 2010)
Renewable Fuel Standard Program http://www.epa.gov/otaq/fuels/renewablefuels/index.htm (accessed 28 May 2010)

1.3. NON-ETHANOL FUELS

AMEC (2007) AMEC to Design First Commercial Scale Biodiesel Production Plant in Canada for Biox Corporation in Hamilton, Ontario, http://www.amec.com/page.aspx?pointerid=c2674000180443l0a1c5ce27a03f0847 (accessed 28 August 2010).
Atsumi, S., and Liao, J.C. (2008a) Directed evolution of *Methanococcus jannaschii* citramalate synthase for biosynthesis of 1-propanol and 1-butanol by *Escherichia coli*. *Appl. Environ. Microbiol.*, **74** (24), 7802–7808.
Atsumi, S., and Liao, J.C. (2008b) Metabolic engineering for advanced biofuels production from *Escherichia coli*. *Curr. Opin. Biotechnol.*, **19** (5), 414–419.
Atsumi, S., Higashide, W., and Liao, J.C. (2009) Direct photosynthetic recycling of carbon dioxide to isobutyraldehyde. *Nat. Biotechnol.*, **27** (12), 1177–1180.
Bermejo, L.L., Welker, N.E., and Papoutsakis, E.T. (1998) Expression of

clostridium acetobutylicum ATCC 824 genes in *Escherichia coli* for acetone production and acetate detoxification. *Appl. Environ. Microbiol.*, **64** (3), 1079–1085.

DARPA (2010) Justification book volume 1: research, development, test & evaluation, defense-wide–0400, http://www.darpa.mil/Docs/FY2011PresBudget28Jan10%20Final.pdf (accessed 28 May 2010).

Eurostat (2005) Consumption of vehicle fuels in the EU-27, epp.eurostat.ec.europa.eu (accessed 28 May 2010).

Global Bioenergies (2010) www.global-bioenergies.com (accessed 28 May 2010).

Hanai, T., Atsumi, S., and Jiao, J.C. (2007) Engineered synthetic pathway for isopropanol production in *Escherichia coli*. *Appl. Environ. Microbiol.*, **73** (24), 7814–7818.

Hill, J., Nelson, E., Tilman, D., Polasky, S., and Tiffany, D. (2006) Environmental, economic, and energetic costs and benefits of biodiesel and ethanol biofuels. *Proc. Natl. Acad. Sci. U.S.A.*, **103** (30), 11206–11210.

Kalscheuer, R., Stolting, T., and Steinbüchel, A. (2006) Microdiesel: *Escherichia coli* engineered for fuel production. *Microbiology*, **152** (9), 2529–2536.

Keasling, J.S., and Chou, H. (2009) Metabolic engineering delivers next-generation biofuels. *Nat. Biotechnol.*, **26**, 298–299.

Khalil, A.S., and Collins, J.J. (2010) Synthetic biology: applications come of age. *Nat. Rev. Genet.*, **11** (5), 367–379.

Kleinov, A., Paligov, J., Vrbov, M., Mikulec, J., and Cvengro, J. (2007) Cold flow properties of fatty esters. *Process Saf. Environ. Prot.*, **85** (B5), 390–395.

Lee, S.Y., Park, J.H., Jang, S.H., Nielsen, L.K., Kim, J., and Jung, K.S. (2008) Fermentative butanol production by clostridia. *Biotechnol. Bioeng.*, **101** (2), 209–228.

Lestari, S., Mäki-Arvela, P., Beltramini, J., Lu, G.Q., and Murzin, D.Y. (2009) Transforming triglycerides and fatty acids into biofuels. *ChemSusChem*, **2** (12), 1109–1119.

OECD/IEA (2008) Energy technology perspectives. Scenarios and strategies to 2050. Paris. http://www.iea.org/techno/etp (accessed 28 August 2010).

Steen, E.J., Kang, Y., Bokinsky, G., Hu, Z., Schirmer, A., McClure, A., Del Cardayre, S.B., and Keasling, J.D. (2010) Microbial production of fatty-acid-derived fuels and chemicals from plant biomass. *Nature*, **463** (7280), 559–562.

Solomon, B.D. (2010) Biofuels and sustainability. *Ann. N. Y. Acad. Sci.*, **1185**, 119–134.

UNEP (2009) Assessing biofuels, http://www.unep.fr/scp/rpanel/pdf/Assessing_Biofuels_Full_Report.pdf (accessed 28 May 2010).

1.4. ALGAE-BASED FUELS

Beer, L.L., Boyd, E.S., Peters, J.W., and Posewitz, M.C. (2009) Engineering algae for biohydrogen and biofuel production. *Curr. Opin. Biotechnol.*, **20** (3), 264–271.

Carbon Trust (2008) Green oil by 2020, http://www.carbontrust.co.uk/news/news/archive/2008/Pages/algae-biofuels-challenge.aspx (accessed 28 May 2010).

Carbon Trust (2010) Algae Biofuels Challenge, http://www.carbontrust.co.uk/emerging-technologies/current-focus-areas/algae-biofuels-challenge/pages/algae-biofuels-challenge.aspx (accessed 28 May 2010).

Chisti, Y. (2008) Biodiesel from microalgae beats bioethanol. *Trends Biotechnol.*, **26** (3), 126–131.

Dasolar Energy (2010) http://www.dasolar.com/alternative-energy/biofuels (accessed 28 May 2010).

Hu, Q., Sommerfeld, M., Jarvis, E., Ghirardi, M., Posewitz, M., Seibert, M., and Darzins, A. (2008) Microalgal triacylglycerols as feedstocks for biofuel production: perspectives and advances. *Plant J.*, **54**, 621–639.

Pulz, O. (2001) Photobioreactors: production systems for phototrophic microorganisms. *Appl. Microbiol. Biotechnol.*, **57**, 287–293.

Richmond, A. (2004) *Handbook of Microalgal Culture*, Blackwell Science, Oxford.

Smith, V.H., Sturm, B.S., deNoyelles, F.J., and Billings, S.A. (2009) The ecology of algal biodiesel production. *Trends Ecol. Evol.*, **25** (5), 301–309.

Solazyme (2010) Technology, http://www.solazyme.com/content/technology (accessed 2 September 2010).

Spolaore, P., Joannis-Cassan, C., Duran, E., and Isambert, A. (2006) Commercial applications of microalgae. *J. Biosci. Bioeng.*, **102**, 87–96.

Stephens, E., Ross, I.L., King, Z., Mussgnug, J.H., Kruse, O., Posten, C., Borowitzka, M.A., and Hankamer, B. (2010) An economic and technical evaluation of microalgal biofuels. *Nat. Biotechnol.*, **28** (2), 126–128.

Tredici, M.R. (1999) Photobioreactors, in *Encyclopedia of Bioprocess Technology: Fermentation, Biocatalysis and Bioseparation* (ed. M.C. Flickinger, S.W. Drew), John Wiley & Sons, Inc., New York, pp. 395–419.

Ugwu, C.U., Aoyagi, H., and Uchiyama, H. (2008) Photobioreactors for mass cultivation of algae. *Bioresour. Technol.*, **99**, 4021–4028.

UNEP (2009) Assessing biofuels, http://www.uncp.fr/scp/rpanel/pdf/Assessing_Biofuels_Full_Report.pdf (accessed 28 May 2010).

Waltz, E. (2009) Biotech's green gold? *Nat. Biotechnol.*, **27** (1), 15–18.

Interesting links:

Algal biomass organization annual report. http://www.algalbiomass.org/ (accessed 28 May 2010)

European algae biomass association. http://eaba-association.eu/legislation.php (accessed 28 May 2010)

OriginOil. http://www.originoil.com/ (accessed 28 May 2010)

1.5. HYDROGEN PRODUCTION

Amos, W.A. (2004) Updated cost analysis of photobiological hydrogen production from chlamydomonas reinhardtii green algae. Milestone completion report. http://www.nrel.gov/docs/fy04osti/35593.pdf (accessed 28 May 2010).

Garman, D., Eiler, J.E., Tromp, T.K., Shia, R.L., Allen, M., Yung, Y.L., Keith, D.K., and Farrell, A.E. (2003) The Bush administration and hydrogen. *Science*, **302** (5649), 1331.

Ghirardi, M.L., Dubini, A., Yu, J., and Maness, P.C. (2009) Photobiological hydrogen-producing systems. *Chem. Soc. Rev.*, **38**, 52–61.

Hemschemeier, A., and Happe, T. (2005) The exceptional photofermentative hydrogen metabolism of the green alga *Chlamydomonas reinhardtii. Biochem. Soc. Trans.*, **33**, 39–41. (accessed 28 May 2010).

HTAC (2010) 2009 HTAC annual report: the state of hydrogen and fuel cell commercialization and tech development, http://www.hydrogen.energy.gov/pdfs/2009_htac_annual_report.pdf (accessed 28 May 2010).

Hydrogen Association (2010) Hydrogen safety, http://www.hydrogenassociation.org/general/factSheet_safety.pdf (accessed 28 May 2010).

Jacobson, M.Z., Colella, W.G., and Golden, D.M. (2005) Cleaning the air and improving health with hydrogen fuel-cell vehicles. *Science*, **308**, 1901–1905.

Kammcn, D.E., Lipman, T.E., Lovins, A.B., Lehman, P.A., Eiler, J.M., Tromp, J.K., Shia, R.L., Allen, M., and Yung, Y.L. (2003) Assessing the future hydrogen economy. *Science*, **302** (5643), 226b.

Melis, A. (2007) Photosynthetic H2 metabolism in chlamydomonas reinhardtii (unicellular green algae). *Planta*, **226**, 1075–1086.

Melis, A., and Happe, T. (2002) Hydrogen production: green algae as a source of energy. *Plant Physiol.*, **127**, 740–748.

NAE (2004) The hydrogen economy: opportunities, costs, barriers, and R&D needs, http://www.nap.edu/openbook.php?record_id=10922&page=R1 (accessed 28 May 2010).

Pielke, R.A., Jr., Klein, R., Maricle, G., and Chase, T. (2005) Hydrogen cars and water vapor. *Science*, **302** (5631), 1329.

Rifkin, J. (2002) *The Hydrogen Economy*, Tarcher/Putnam, New York.

Schultz, M.G., Diehl, T., Brasseur, G.P., and Zittel, W. (2003) Air pollution and climate-forcing impacts of a global hydrogen economy. *Science*, **302**, 624–627.

Schütz, K., Happe, T., Troshina, O., Lindblad, P., Leitão, E., Oliveira, P., and Tamagnini, P. (2004) Cyanobacterial H2 production: a comparative analysis. *Planta*, **218**, 350–359.

Shere (1937) Hindenberg disaster, http://iconicphotos.wordpress.com/2009/07/25/hindenberg-disaster/ (accessed 28 May 2010).

Stripp, S.T., Goldet, G., Brandmayr, C., Sanganas, O., Vincent, K.A., Haumann, M., Armstrong, F.A., and Happe, T. (2009) How oxygen attacks [FeFe] hydrogenases from photosynthetic organisms. *Proc. Natl. Acad. Sci. U.S.A.*, **106**, 17331–17336.

Surzycki, R., Cournac, L., Peltier, G., and Rochaix, J.D. (2007) Potential for hydrogen production with inducible chloroplast gene expression in *Chlamydomonas*. *Proc. Natl. Acad. Sci. U.S.A.*, **104** (44), 17548–17553.

Tamagnini, P., Axelsson, R., Lindberg, P., Oxelfelt, F., Wünschiers, R., and Lindblad, P. (2002) Hydrogenases and hydrogen metabolism of cyanobacteria. *Microbiol. Mol. Biol. Rev.*, **66**, 1–20.

Tetali, S.D., Mitra, M., and Melis, A. (2007) Development of the light-harvesting chlorophyll antenna in the green alga *Chlamydomonas reinhardtii* is regulated by the novel tla1 gene. *Planta*, **225** (4), 813–829.

Tromp, T.K., Shia, R.L., Allen, M., Eiler, J.M., and Yung, Y.L. (2003) Potential environmental impact of a hydrogen economy on the stratosphere. *Science*, **300** (5626), 1740–1742.

US DOE (2010) Fuel cell technology validation, http://www1.eere.energy.gov/hydrogenandfuelcells/tech_validation (accessed 28 May 2010).

Zeman, F.S., and Keith, D.W. (2008) Carbon neutral hydrocarbons. *Philos. Transact. A Math. Phys. Eng. Sci.*, **366** (1882), 3901–3918.

Links to Hydrogen Safety pages:

Hydrogen and fuel cell safety: www.HydrogenSafety.info (accessed 28 May 2010)

Hydrogen Safety Bibliographic Database: www.hydrogen.energy.gov (accessed 28 May 2010)

Hydrogen Safety for First Responders: http://www.hydrogen.energy.gov/firstresponders.html (accessed 28 May 2010)

Reporting Lessons Learned: www.h2incidents.org (accessed 28 May 2010)

Sharing Best Practices: www.h2bestpractices.org (accessed 28 May 2010)

Other interesting links:

Anastasios Melis http://epmb.berkeley.edu/facPage/dispFP.php?I=25 (accessed 28 May 2010)

Biological hydrogen production http://en.wikipedia.org/wiki/Biological_hydrogen_production (accessed 28 May 2010)

Engineered Modular Bacterial Photoproduction of Hydrogen http://biomodularh2.epigenomique.genopole.fr/index.php/Public/Impact (accessed 28 May 2010)

Growing hydrogen for the cars of tomorrow http://www.science.org.au/nova/newscientist/111ns_002.htm (accessed 28 May 2010)

1.6. MICROBIAL FUEL CELL (MFC) AND BIO-PHOTOVOLTAIC

Bullen, R.A., Arnot, T.C., Lakeman, J.B., and Walsh, F.C. (2007) Biofuel cells and their development. *Biosens. Bioelectron.*, **21** (11), 2015–2045.

Cao, X., Huang, X., Liang, P., Xiao, K., Zhou, Y., Zhang, X., and Logan, B.E. (2009) A new method for water desalination using microbial desalination cells. *Environ. Sci. Technol.*, **43** (18), 7148–7152.

DARPA (2010) Benthic unattended generator, http://www.nrl.navy.mil/code6900/bug/ (accessed 28 May 2010).

Davis, F., and Higson, S.P.J. (2007) Biofuel cells – Recent advances and applications. *Biosens. Bioelectron.*, **22** (7), 1224–1235.

De Schamphelaire, L., Van den Bossche, L., Son Dang, H., Höfte, M., Boon, N., Rabaey, K., and Verstraete, W. (2008) Microbial fuel cells generating electricity from rhizodeposits of rice plants. *Environ. Sci. Technol.*, **42** (8), 3053–3058.

Debabov, V.G. (2008) Electricity from microorganisms. *Mikrobiologiia*, **77** (2), 149–157. (in Russian).

Donovan, C., Dewan, A., Heo, D., and Beyenal, F. (2008) Batteryless, wireless sensor powered by a sediment microbial fuel cell. *Environ. Sci. Technol.*, **42** (22), 8591–8596.

Ghangrekar, M.M., and Shinde, V.B. (2008) Simultaneous sewage treatment and electricity generation in membrane-less microbial fuel cell. *Water Sci. Technol.*, **58**, 37–43.

GoPubMed (2010) Pubmed metasearch website www.gopubmed.org (accessed 28 August 2010).

Logan, B. (2009) Scaling up microbial fuel cells and other bioelectrochemical systems. *Appl. Microbiol. Biotechnol.*, **85** (6), 1665–1671.

Logan, B., Aelterman, P., Hamelers, B., Rozendal, R., Schroder, U., Keller, J., Freguia, S., Verstraete, W., and Rabaey, K. (2006) Microbial fuel cells: methodology and technology. *Environ. Sci. Technol.*, **40** (17), 5181–5192.

Lovley, D.R. (2006) Bug juice: harvesting electricity with microorganisms. *Nat. Rev. Microbiol.*, **4**, 467–508.

Lovley, D.R. (2008) The microbe electric: conversion of organic matter to electricity. *Curr. Opin. Biotechnol.*, **19**, 1–8.

Makezine (2006) How to: make microbial fuel cells, http://blog.makezine.com/archive/2006/05/how_to_make_microbial_fue.html (accessed 28 May 2010).

Pant, P., Van Bogaert, G., Diels, L., and Vanbroekhoven, K. (2010) A review of the substrates used in microbial fuel cells (MFCs) for sustainable energy production. *Bioresour. Technol.*, **101** (6), 1533–1543.

Peter, L.M., Walker, A.B., Boschloo, G., and Hagfeldt, A. (2006) Interpretation of apparent activation energies for electron transport in dye-sensitized nanocrystalline solar cells. *J. Phys. Chem. B*, **110** (28), 13694–13699.

Reguera, G., McCarthy, K.D., Mehta, T., Nicoll, J.S., Tuominen, M.T., and Lovley, D.R. (2005) Extracellular electron transfer via microbial nanowires. *Nature*, **435**, 1098–1101.

Ren, Z., Steinberg, L.M., and Regan, J.M. (2008) Electricity production and microbial biofilm characterization in cellulose-fed microbial fuel cells. *Water Sci. Technol.*, **58** (3), 617–622.

Shantaram, A., Beyenal, H., Raajan, R., Veluchamy, A., and Lewandowski, Z. (2005) Wireless sensors powered by microbial fuel cells. *Environ. Sci. Technol.*, **39** (13), 5037–5042.

Tender, L.M., Reimers, C.E., and Stecher, H.A. (2002) Harnessing microbially generated power on the seafloor. *Nat. Biotechnol.*, **20** (8), 821–825.

Wang, X., Feng, Y.J., and Lee, H. (2008) Electricity production from beer brewery wastewater using single chamber microbial fuel cell. *Water Sci. Technol.*, **57** (7), 1117–1121.

Interesting links:

Microbial fuel cells. http://www.microbialfuelcell.org/ (accessed 28 May 2010)

2
Bioremediation

Ismail Mahmutoglu, Lei Pei, Manuel Porcar, Rachel Armstrong, and Mark Bedau

2.1
Bioremediation in General

2.1.1
Introduction

Pollution of soils is caused mainly by human activities – not only found in the soils near the mines or industrial sites, but also in arable lands due to the intensive agriculture practices. To date, the contamination of ground- and surface waters constitute a major threat to public health in developing and industrial countries, including many European countries, Bangladesh, China and the United States. Two major types of pollutants can be identified: xenobiotics (with a considerable proportion of pesticides) and a variety of minerals. The latter category includes minerals that are not pollutants at low concentration, such as metal ions, iron included, phosphate or nitrate or ammonium, and minerals that are always toxic such as chromium, cadmium, mercury, arsenic and so on. Xenobiotics are most often aliphatics and aromatics and, to a lesser extent, sulfur-containing molecules or more exotic compounds. It was initially expected that aliphatics and aromatics could be readily amenable to degradation using microbes, either in pure or in mixed cultures. We have had to lower our expectations in this respect. This could have been foreseen by witnessing the stability of wood, for example, which has a large lignin component, whose basis is aromatics (typically derivatives of coumarin and a variety of methoxylated phenylpropanoids).

Technically, SB could make use of existing pathways leading to the degradation of complex alcohols, putting them together in synthetic pathways and combining microorganisms. However, a significant number of aromatics have the property of either disrupting membranes or permeating through them, causing the collapse of the electrochemical potential that maintains viability. This implies that, even when successful pathways are put together, they will only operate at low pollutant concentrations. Optimally using SB approaches will therefore require combining bioremediation with physico-chemical approaches that can promote decontamination

Synthetic Biology: Industrial and Environmental Applications, First Edition. Edited by Markus Schmidt.

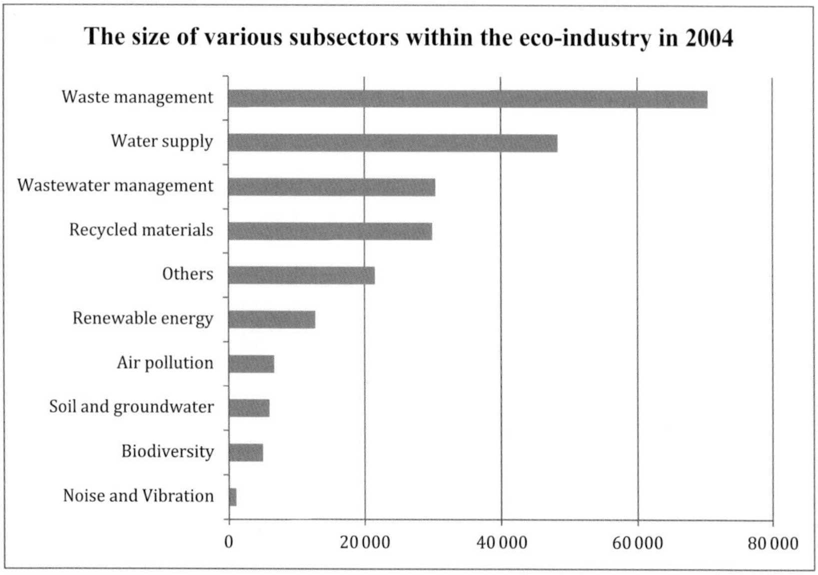

Figure 2.1 The turnover of the European eco-industry in 2004, including subsectors (in Mio. €).

of fields. Another clear technical challenge is common to most SB applications when they are supposed to operate on a large scale: scaling up living organisms parallels evolution. Unless we are able to harness evolutionary trends to serve human goals, the chosen microbes may well rapidly produce divergent organisms that can multiply without performing the desired tasks.

A specific situation is encountered with soils polluted by radioactive material. Here, as for toxic minerals, the idea is to construct, using SB, organisms that concentrate and precipitate the toxic material.

2.1.2
Economic Potential

The economic potential is huge. It is proportional to the surface of land occupied by highly polluting material, to the amount of waste and to the volume of polluted water. The eco-industry of the EU27 in Europe had a turnover of €232 billion (2.2% of GDP) in 2004. Corrected for inflation, the annual growth rate is 5.9%. As Figure 2.1 shows, the four largest subsectors take up to roughly three-quarters of the total eco-industry, with waste management (30%) as the largest one followed by water supply (21%), wastewater management (13%) and recycled materials (13%).[1]

1) http://ec.europa.eu/enterprise/newsroom/cf/document.cfm?action=display&doc_id=5416&userservice_id=1

In the case of wastewater treatment (annual turnover around €30 billion), SB can make many significant contributions. It can help to enhance the effectiveness of the bioprocess, reducing the energy requirements – a key economic and environmental factor in wastewater treatment.

The turnover of the construction-based bioremediation industry in Europe, which is practically overlapping with the definition of soil and groundwater, is estimated to be around €6 billion/year. The expectation is that SB methods will increase effectiveness, reducing costs and encouraging more investors to purchase brownfields. This, in turn, would increase overall turnover. One aspect has to be underlined: markets depend largely on the legal framework in a country, that is, whether or not remediation is legally required. Also, action has to be taken only if an immediate danger is foreseeable – a contaminated soil that is contained and not connected to the groundwater does not have to be cleaned, therefore limiting the legally mandated use of bioremediation techniques.

2.1.3 Environmental Impact

The impact of nitrate, beside its role in water pollution, is considerable. The microbial flora of soils converts a very significant proportion of the ion into nitrous oxide, which is a GHG with a potential 300 larger than that of carbon dioxide for equivalent mass. The contribution of pig manure, for example, is enormous in Europe (Denmark and France in particular) and increasing extremely fast in China (http://bioinf.gen.tcd.ie/EAGLES/en/program.html). For other minerals such as metal ions, arsenic or uranium, biomineralization might be the best option. Here, microbes would concentrate pollutants and excrete them as insoluble materials. This is the exact opposite of those microbial processes used to leach interesting metals out of ores that have only a low level of that metal, for example, copper.

SB can also help eliminate the so-called endocrine substances such as hormones, pharmaceutically active substances, flexibilisers or X-ray contrast media. They skip through the treatment plant due to their persistency or their very low concentrations (nano- to picomolar range), which are often still high enough to negatively affect aquatic life and human beings (BAFU, 2009). They could be eliminated by an after-treatment with highly specific organisms or bio-devices.

Another major problem is the dredge sludge from European rivers and canals, which adds up to millions of tonnes each year. This sludge cannot be dumped on regular soil landfill due to its high organic content and sometimes contaminations. Present biological systems are not very suitable due to the high content of water and fine constituents. Here, innovative systems created with the aid of SB could also help to solve this waste problem.

The situation with xenobiotics has been explored by a large number of studies. About two decades ago, the hope was that simple genetic engineering would permit easy degradation of aromatics. Unfortunately this proved to be incorrect, and we are still very far from succeeding in degrading xenobiotics, aromatics in particular. DDT remains a stable contaminant despite considerable research for several decades.

SB would benefit from genomic studies of organisms that survive in highly polluted environments. This approach would help to identify the functions that microbes use to survive.

2.1.4 Foreseeable Social and Ethical Aspects

Bioremediation is seen as positive. The benefits are obvious: new lands become open to agriculture, the quality of water is improved and so on. In practice, bioremediation is not new, neither for water nor for soil decontamination. More easily than remediation, bioremediation can be directly associated (in time and space) with production.

Different strategies – which raise different risks – have to be distinguished. This means that the level of public acceptance will also differ. Bioremediation can be done with specially designed microorganisms and plants, or by adapting already existing microorganisms and plants to this task. In the latter case, these organisms can be directly used or modified to better adapt them to specific environments and tasks. Bioremediation can degrade toxic substances (Shimao, 2001; Cao, Nagarajan, and Loh, 2009; Rojo, 2009) or accumulate them within the organisms (Kavamura and Esposito, 2010).

The ethical issues are first linked with the use of "synthetic" organisms. The risks are diverse, the first being that the operations will typically be done in an open space, with no restrictions on the proliferation capacities of the organisms. Microorganisms used for decontamination might favorably compete with preexisting organisms, in particular at the boundaries of the decontaminated zones, alter the ecosystems and reduce biodiversity.

Biosecurity mainly concerns the processes in which the toxic substances are accumulated in organisms. These toxic wastes might be directly used by humans or exploited as a source of material (e.g., uranium) for the production of bioweapons.

Most such processes can be used in any country, regardless of its economic level. Their development, however, will require sophisticated technologies, especially when a community of organisms is involved. This level of sophistication goes far beyond the capacity of "garage" amateurs, thus the immediate biosecurity risk of amateur biologists is rather low.

2.2 Detection of Environmental Pollutants (Biosensors)

2.2.1 Introduction

A biosensor is commonly referred to a biological based device that combines a receptor component to capture a target molecule and a signal transduction component to convert the target–receptor binding event into measurable signals. Such measurable signals can be in either form of fluorescence, chemiluminescence, colorimetric, electrochemical and magnetic responses that produce a proportional signal to detect molecules of interest. The biosensors provide a simple yet reliable measurament to detect agents of interest in the environment. The practical applications are medical diagnostics, environmental monitoring, food safety and military applications. Compared to the standard analytical methods that directly monitor environmental pollutants in the field, biosensors offer the distinct advantages of being cost-efficient and quantitative without complex sample preparation.

Synthetic biology gave a boost to the biosensors field through the emergence of new developments from systemic design, novel means of signal transduction and control elements (e.g., well defined genetic circuits for logical gates). Many new concepts were introduced with the advent of SB (Nakamura, Shimomura-Shimizu, and Karube, 2008). In conjunction with advanced detection technologies, biosensors are becoming important tools for environmental monitoring, where rapid and remote monitoring can identify environmental hazards ranging from heavy metals, insecticides, genotoxins, phenols and arsenic to trinitrotoluene (TNT). There are four types of biosensors for environmental monitoring.

1) Enzyme biosensors are widely used to monitor environmentally important substrates based on the principle of measuring the inhibition of a specific activity of the enzyme due to the presence of the target substrate. The enzymes commonly used are oxido-reductases and hydrolases.

2) Immunobiosensors are devices whose elements are based on the interaction between antibodies and antigens.

3) Nucleic acid-based biosensors employ oligonucleotides as the sensing elements, with a known sequence of bases, or a complex structure of DNA or RNA. Nucleic acid biosensors can be used to detect DNA/RNA fragments or other chemical substances. They have been used to detect environmental pollution and toxic substances. In some applications, DNA/RNA is the target analyte and it is detected through the hybridization reaction. In the other applications, DNA/RNA plays the role of a receptor of specific biological or chemical substances, such as DNA-binding proteins, pollutants or drugs targeting nucleic acids.

   - DNA biosensors are based on the affinity of the target molecules (particularly those of mutagenic and carcinogenic activity) to immobilized DNA, or the *in situ* hybridization of the target DNA to immobilized ones.

   - RNA sensors are parts that detect signals. RNA sensors detect diverse signals, such as temperature and molecular ligands, through various binding events, including hybridization and tertiary interactions. The binding event encoded in an RNA sensor is generally transduced to an actionable event such that RNA sensors are typically coupled to other RNA parts (Figure 2.2).

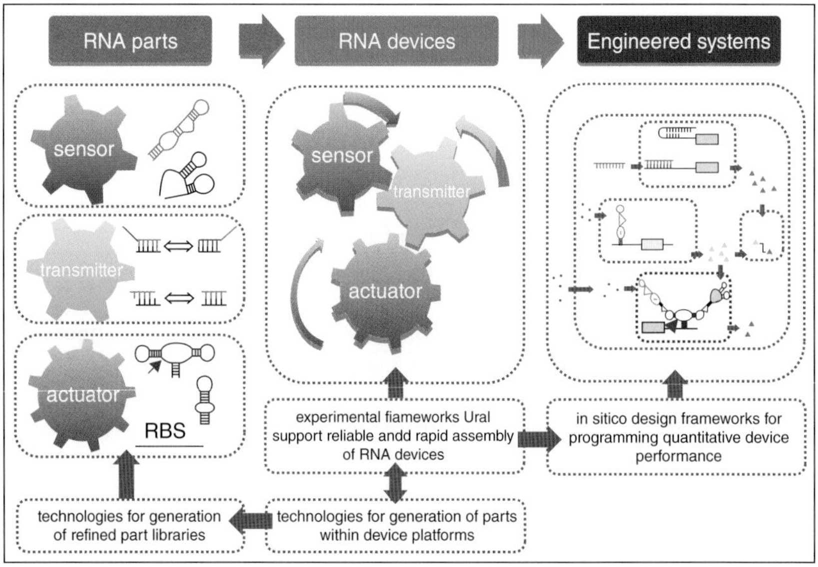

Figure 2.2 Process flow of enabling technologies supporting device design and implementation into RNA sensors (Win, Liang, and Smolke, 2009).

4) Whole-cell biosensors, particularly those using microbes, are analytical devices that couple the whole cell (microorganism) with a transducer. Earlier microbial biosensors were designed to detect a substrate or an inhibitor of the microbial pathways of respiratory and metabolic functions. The recent biosensors are based on fusing reporter elements (such as *lux*, *gfp*, or *lacZ*) to an inducible gene promoter to detect targets of interest (Figure 2.3). This type of biosensor can be applied to assess biological effects such as biological oxygen demand (BOD), toxicity, ecotoxicity or estrogenicity. Four types of synthetic biosensor have been proposed by Khalil and Collins (2010).

   - Transcriptional biosensors are built by linking the quorum sensing like promoters to gene circuits for programmed transcriptional changes. A transcriptional gate is designed to sense and report only the simultaneous presence of environmental signals (e.g., salicylate and arabinose).
   - Translational biosensors are built by coupling RNA aptamer domains to RNA regulatory domains. The target molecule will be recognized and bound by the aptamer stem of the RNA biosensor. This will lead to a conformational change in the molecule and inhibit the translation of an output reporter.
   - Post-translational biosensors are protein-based biosensors. This type of sensors is designed based on the principle that signal transduction pathways are organized on essentially hierarchical schemes based on sensitive

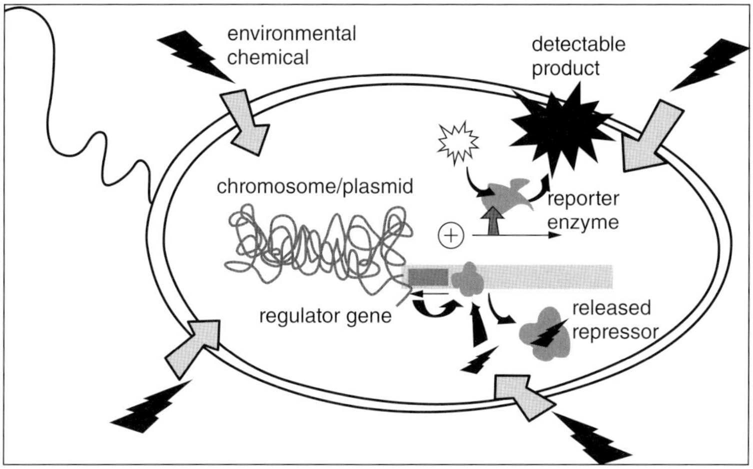

Figure 2.3 Schematic representation of a bacterial biosensor for an environmental chemical (Harms, Wells, and van der Meer, 2006).

elements and downstream transducer modules. Therefore, the protein receptor of a pathway and the associated cascade are excellent elements for signal detection.

– The hybrid type of sensors is the combination of a post-translational circuit with a transcriptional one.

2.2.2 Economic Potential

The global market size for biosensors was about $7.3 billion in 2003, with a predicted growth rate of 10.4% (Fuji-Keizai Report, 2004; Luong, Male, and Glennon, 2008). According to the market report by Frost and Sullivan, the world biosensor market comprises six vertical markets: home diagnostics, point of care, research laboratories, process industries, environmental and bio-defense applications. The biosensor market for environmental monitoring is relatively small (<15% of global revenues). Ideal biosensors should be fast, easy to use, specific and inexpensive. Significant upfront investment required for developing biosensors is a key challenge faced by manufacturers. In addition, prolonged R&D timelines make it difficult to justify the high costs in the absence of huge volume markets. Smaller suppliers in particular are less likely to obtain sufficient funding. A significant upfront investment in R&D is a prerequisite for the commercialization of biosensors (Harms, Wells, and van der Meer, 2006). There are 21 companies that produce biosensors for environmental applications in the EU, which is about 1.4% of all companies producing environmental tests in Europe (BIO4EU, 2007).

Besides the market value of biosensors, these devices entail economic impacts. One such impact is the development of harmful algal bloom (HAB) biosensors (ACT, 2002). HABs have a significant impact on marine resources, local economies and public health in most coastal nations. The economic loss caused by HABs is approximately $40 million annually. Several technological advances have been made in the detection of algal species of HAB and toxins. One specific area of HAB sensor technology advancement is molecular-based detection: molecular-based detection of individual HAB species and species-specific probes for numerous toxins commonly found in water. Such probes encompass both antibodies targeting cell surface antigens and nucleic acid probes detecting intracellular genetic signatures.

2.2.3
Environmental Impact

The development of cost-effective on-site methods for environmental monitoring is critical for managing environmental contamination. Biosensors show dual functions that can complement both laboratory-based and field analytical methods for environmental monitoring. A wide range of biosensors have been reported for potential environmental applications.

Whole-cell biosensors are designed to respond to specific signals from the environment. They can be deployed to recognize pathogenic microbes in water and food, such as *Salmonella* or *Escherichia coli*, or to detect hazardous chemicals in soil, air or water. Homme Hellinga of Duke University and colleagues proposed a computational design of receptor and sensor proteins with novel functions such as to bind targeted chemicals (Looger *et al.*, 2003). They reengineered a sugar-binding protein of *E. coli* to bind to the molecules of explosive TNT. The redesigned protein was plugged into an engineered gene circuit, which was then integrated into a bacterium to create a whole-cell sensor that turns green when it senses its target chemical. Similar microbial biosensors could be designed to detect environmental pollutants. Using a synthetic biology approach, a team at Emory University in Atlanta has now equipped an engineered strain of *E. coli* with the ability to hunt down the herbicide atrazine and metabolize it (Sinha *et al.*, 2010). Atrazine is an environmental pollutant that can be harmful to wildlife. Key to the transformation is the combination of a synthetic switch that allows the bacterium to chase the chemical and a gene taken from another species of bacterium for breaking down atrazine. Justin Gallivan and his team have used RNA to develop an atrazine-binding molecule (a synthetic riboswitch). It is a piece of RNA that binds to a small molecule and changes shape in doing so. This then alters gene expression. In the second step, the switch-carrying bacteria have been equipped with an atrazine-degrading gene isolated from a different bacterium species. The resulting bacteria demonstrate their seek and destroy behavior by forming rings in petri dishes covered with atrazine as they move toward the atrazine and clear it from the plate. A bacterial arsenic biosensor was an iGEM project accomplished in the University of Edinburgh in 2006 (Team Edinburgh iGEM,

2006). The Edinburgh team proposed to develop a device that responds to a range of arsenic concentrations and produces a change in pH that can be calibrated in relation with the arsenic concentration. They coupled a pH change, an easily measurable parameter to the presence of arsenic. Compared to previous designs of arsenic detectors, the signal was easily detectable using a pH meter. It was also relatively easy to make the bacterial biosensors available to detect arsenic in rural areas. Besides arsenic, biosensors have been developed to detect environmentally significant metal ions, primarily using enzymes or whole-cell sensors. A urease enzyme-based fiber optic biosensor can detect a number of metal ions by measuring the inhibited degree of the response of the biosensor. For example, biosensor response is most sensitive to Hg^{2+}, with a detection limit of 10 nM. In whole-cell biosensors, engineered bacteria whose genes are responsible for Hg^{2+} detoxification are linked to light-emitting genes. The biosensor response is typically specific to Hg^{2+}, with detection limits in the low nanomolar range (Tescione and Belfort, 1993).

Phenol compounds have been spread widely and are listed in the Priority Substances List (http://ec.europa.eu/environment/water/water-framework/priority_substances.htm). The list also contains the pesticides that should be main target for the development of environmental biosensors. Pesticides function by means of interacting with a specific biochemical target either as a substrate or as an inhibitor. The Biosensors Research Group at the EPA's National Exposure Research Laboratory–Las Vegas is conducting research to develop electrochemical biosensors to detect phenols and organophosphate (OP) pesticides (Biosensor at EPA, 2009). Reports have shown the tyrosinase enzyme electrode to be sensitive and durable in measuring phenols in water and soil. The challenges to develop detection devices involve the development of more robust techniques for immobilization of the enzyme to the devices. Research conducted by the EPA has focused on developing biosensors for environmentally significant phenols from groundwater, soils and sludges. EPA's other research activities aim to develop biosensors that can be used to detect OP pesticides based on enzyme systems. Among them, organophosphorus hydrolase (OPH) and acetylcholinesterase have been studied. The OPH-based method for measuring OP pesticides is based on a substrate-dependent change in pH near the enzyme. The pH change is monitored using fluorescein isothiocyanate (FITC), which is covalently immobilized to the enzyme. They developed a method that employs the use of polymethylmethacrylate beads to which the FITC-labeled enzyme is adsorbed. The analytes are then measured using a microbead fluorescence analyzer.

Biosensors for Effective Environmental Protection and Commercialization (Beep-c-en, 2010), a project funded by the European Commission within Call FP 7-SME-2008-01, deals with the integration of innovative biosensor research and technology and their exploitation by industry and/or other socio-economic entities in the fields of environment and agro-industry. The project aims at building up a biosensor modular industrial platform that can be easily adopted to detect pesticides, heavy metals and organic compounds in water. One task of the project is to build three types of sensors, namely: MultiLights, MultiAmps and MultiTasks.

These sensors will be assembled in a full system platform characterized by high specificity, high selectivity, high stability, high sensitivity, room temperature operation, simple use, low cost, fast response time, minimum sample pretreatment, small dimension and ease of transport for *in situ* measurements, real time online measurements and easy interface with integrated circuits, enabling cost-effective manufacturing.

2.2.4 Foreseeable Social and Ethical Aspects

The principle is to use the extraordinary capacities of organisms to respond to "weak" signals and to discriminate between highly similar molecules. A good example is olfaction. The final goal – to combat pollution – will find easy acceptance. The use of organisms to detect pollution is not new. Birds, for example, were used in coalmines to detect the presence of toxic gases. More recently, but before the advent of synthetic biology, organisms were modified to change their color, or to emit light, in conditions of environmental stress.

The exact future course (Looger *et al.*, 2003; Li *et al.*, 2010) and the ethical issues remain fuzzy. The nature of the pollutants and the conditions in which they will be detected must be more precisely known for the risks to be estimated. It is not obvious that synthetic organisms will favorably compete for these tasks with more traditional chemical or physical methods. It is also possible to imagine that these new biosensors will not be made of living organisms, but of molecular nanomachines extracted from these organisms (Ehrentreich-Förster *et al.*, 2008).

The risk might consist in the dissemination of living biosensors. It might be particularly high if they are used to monitor water quality. The issue of proper containment of the leasing of synthetic organisms will need serious consideration. Whether these new methods to detect pollution will lead to a democratization and extension of pollutant detection at the global level remains an open question.

Over the last decades, public concern over contamination with persistent pollutants has grown. Every year, hundreds of chemicals of known toxicity that affect the quality of the air, land and groundwater are released into the environment. Although considerable progress has been achieved in the clean-up of polluted soil, sediments and groundwaters during recent years, conventional technologies are economically and environmentally costly. Moreover, it is not always easy to detect toxic substances until their concentrations reach unacceptable levels. The early detection and elimination of pollutants from the environment is essential for the sustainable development of advanced societies, and this objective will be better achieved by using improved biosensors obtained by synthetic biology approaches. This will have an ecological effect (better monitored and protected environments) as well as economic advantages.

Based on the current development of the technology, the applications are not expected to significantly change social interactions in society.

Ethical issues. No additional ethical concern can be envisaged, except for those that apply to any engineering approach to biology, such as the intentional or accidental release of synthetic organisms into the environment, which can alter the ecological conditions of the environment. Another aspect is the misuse of the new synthetic organisms, although such misuse to create biological weapons seems unlikely.

Justice of distribution. Provided that any potential contaminant activity, in any developed or developing country, is strictly regulated to enforce environmental monitoring and protection, it will be cheaper to use improved biosensors than using traditional approaches to clean-up the contaminated areas and then pay for the damage. Intellectual property issues and the legal framework of their application will also influence the justice of distribution.

2.3 Water Treatment

2.3.1 Introduction

Water polllution can be human-driven (domestic or industrial wastewater, contaminated groundwater) or natural (heavy metals of geogenic origin). Contaminated groundwater stems from industrial activities in the past and can be considered as a finite problem because intense care is being taken today to protect groundwater from pollution. Furthermore, groundwater contamination is always associated with soil contamination (see Section 2.5: *Soil and groundwater decontamination*). In contrast, sewage and industrial wastewater are inevitable byproducts of human activities. They will accrue indefinitely and in great bulk, even if their extent is reduced by intelligent measures. Solutions have existed for a very long time and treatment plants exist almost everywhere, with considerable success. Their engineering is already well understood and organized, and microorganisms play a considerable role in the quality of the water produced. This essentially corresponds to the bioremediation of organic pollutants and involves situations where the concentration of xenobiotics is relatively low.

By contrast, mineral pollution, sometimes from natural water environments, is more difficult to tackle. For example, elevated arsenic concentrations typically derive from the weathering of arsenic-bearing minerals or from geothermal sources as well as, to a lesser extent, from anthropogenic origins (e.g., smelting and mining industries). Microorganisms are known to influence arsenic geochemistry by their metabolism, which may include reduction (including arsenate respiration), oxidation and/or methylation reactions. These biological activities affect both the speciation and the toxicity of arsenic, the reduced species arsenite As[III] being far more toxic than the oxidized form arsenate As[V].

Finding means of preserving and restoring natural environments constitutes a major challenge facing modern society. Metal-oxidizing or metal-reducing bacteria represent an attractive tool to restore contaminated sites, but they remain fairly poorly characterized, in particular in terms of their metabolic capacities. Some bacteria thriving in such harsh environments have been somewhat characterized at the genome level. They display significant metabolic versatility and the ability to restore life-nurturing conditions in the environment. An important feature is to avoid gasification of arsenic (the same is true for mercury) and to promote its mineralization into insoluble material.

2.3.2
Economic Potential

As mentioned above (Section 2.1.3), the dominant sector in water treatment is wastewater treatment, with a turnover of around €30 billion in 2004 (groundwater treatment including soil remediation around €6 billion). The water treatment technology applied by wastewater treatment plants i(WWTPs) based almost exclusively on biological principles. Due to the high oxygen demand, which requires tremendous amounts of pressurized air, biological water treatment is energy and cost intensive. Thus, any contribution to a better process efficiency will go beyond a direct and substantial economic impact.

An interesting issue is the fermentation of sewage prior to the treatment in the WWTPs. Here, SB could also help to develop new methods to convert the organic freight of sewage into an energy source (e.g., H_2, CH_4). This would yield renewable, CO_2-neutral energy and would, at the same rate, reduce the chemical oxygen demand (COD) and BOD of the wastewater. This would reduce the energy demand of the aerobic treatment process and enhance the biodegradability of the organic freight due to previous breakdown processes in the fermentation phase.

With regard to drinkable water, the question of its availability within the next few years is central to international discussions. Lack of clean water will soon (and probably already does) create tensions that could lead to war. Although water treatment developed long ago, in some situations (such as arsenic-rich water in China or in Bangladesh) the question of water depollution has still remained crucial.

2.3.3
Environmental Impact

Any SB-related improvement of process efficiency in biological water treatment would have both an economic and an environmental effect (reduced energy demand, reduced CO_2 emissions, better effluent quality).

Further environmental benefits of SB could be a better elimination of micropollutants (see also Section 2.1.3), which are difficult to treat with traditional biological, physical, physico-chemical or chemical methods.

SB could also help to develop novel treatment methods for water pollution problems which cannot be solved now with the present state of knowledge.

2.3.4 Foreseeable Social and Ethical Aspects

Water treatment is a huge and global problem (Tal, 2006; Shannon *et al.*, 2008). The lack of clean water resources is at the origin of many conflicts. Contaminated waters generate many diseases (e.g., poisoning), which put a burden on many countries and hamper their development. Any technological progress is welcome in this field.

Water treatment, including the use of natural microbes, is a "natural" process used by human beings for centuries and rationally organized in the late nineteenth century. No clear changes in social interactions can be predicted and more sophisticated technologies linked with synthetic biology should be developed.

The ethical issues are mainly associated with the potential risks. The degree of confinement will be crucial. Water after treatment will be consumed directly by humans. This raises the risk that fears and reactions similar to those observed in the case of GMOs will surge ("no GMO on our plates or in our drinking glasses!").

The treatments might generate non-toxic compounds or lead to the sequestration of the toxic compounds in muds or organisms. Such treatments might therefore generate a novel problem due to the difficulty of the safe storage of wastes.

Water has always been considered as a potential target for terrorism and bioterrorism. The use of "synthetic organisms" in the treatment of water might open new opportunities for smart bioterrorists.

The use of microorganisms in water treatment might reduce the scale at which a treatment is possible, a potential benefit for less developed countries. At the same time, the development of well adapted microorganisms, in particular when a community of microorganosms is required, will necessitate sophisticated molecular technologies, available in a limited number of countries.

2.4 Water Desalination with Biomembranes

2.4.1 Introduction

The increase in the world population and the reduction of freshwater resources due to overuse or contamination are threatening human life on Earth. Millions if not billions of people have no access to clean drinking water resources. Freshwater is also needed for agricultural purposes to satisfy the growing demand for food. Finally, there are indications that climate change may lead to a desertification of certain areas with a high population densities.

One feasible approach to overcome this shortage is the desalination of seawater, which is abundantly present in the world's oceans. Salty groundwater, found in many, especially arid, countries can be used for this purpose too. Today, this

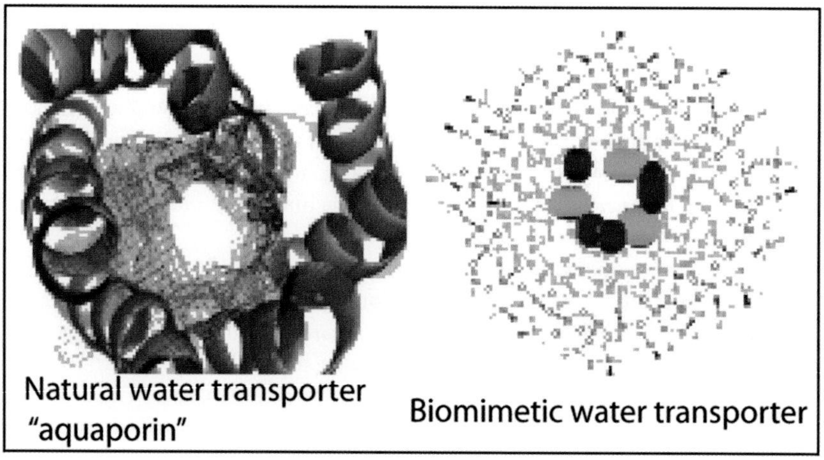

Figure 2.4 A possible molecular model for biological water desalination.

process is carried out by techniques like reverse osmosis (RO) or distillation. Both are very expensive with regard to investment and operation costs.

Nature has developed effective systems to transport ions against concentration gradients. The energy to overcome the osmotic counterforce is provided by biochemical energy carriers (such as ATP) or even directly by sunlight (i.e., the ion pump of Halobacteria). With the aid of SB, tailor-made organisms could be created for technical-scale desalination. These systems could also be adapted to technical devices (membranes, tubes, etc.) for use in solar desalination modules.

Intensive research is being conducted on biological ion pump systems like halorhodopsin, a light-driven ion pump for chloride ions found in the Halobacteria. The structural features and mechanisms of these systems are well investigated. A SB-based application can be expected within the next 10–20 years (Figure 2.4).

2.4.2
Economic Potential

The economic potential of this technology would be enormous. The economical benefits can be considered in two parts.

First, the system would create a new and fast-growing industry (like today's photovoltaic or wind turbine industry) producing the hardware. The lower costs compared to the present systems would greatly enhance the global turnover.

Secondly, the system would greatly stimulate agricultural production in arid areas (which are also present in some parts of Europe). Here, freshwater produced along coastlines could be used locally or transported inland by pipelines. This would help to improve the economy of these regions and would support the battle against hunger.

2.4.3 Environmental Impact

As mentioned above, current desalination systems entail extensive operational costs consisting almost exclusively of energy costs. These processes consume enormous amounts of energy and create CO_2. In optimized currently available systems, the energy demand for the production of $1\,m^3$ freshwater from seawater is around 2–3 kWh.

SB-created systems based on biological principles and driven by solar light would help reduce CO_2 emissions. No waste production is expected. Note, however, one general problem associated with sea water desalination systems: in some cases, the highly concentrated residual salt solution from RO plants had a negative effect on marine life when delivered back close to the desalination plant. This calls for discharging the concentrate in a manner that it can readily be diluted with seawater and does not impact marine life. Beyond such potential local concentration effects, no significant increase in the global salt concentration of seawater is expected because the amounts of freshwater (and of concentrate, respectively) produced are negligible compared to the volume of the oceans.

A significant increase in the production of freshwater would boost plant growth (e.g., in arid areas or deserts), with a positive effect on CO_2 reduction and climate regulation.

SB applications are sustainable because they are generally based on renewable systems. No potential risks are expected when the applications are associated with devices like membranes or tubes. If living organisms are used, however, the risks of a possible release into the environment must be assessed carefully.

2.4.4 Foreseeable Social and Ethical Aspects

Access to drinkable water is a major issue today. New procedures involving the use of microorganisms, less expensive than traditional ones that could be developed on a small scale would be a plus (Tal, 2006; Shannon *et al.*, 2008; Cau *et al.*, 2009). They would help solve conflicts within or between countries.

Treatment of water will take place in confined spaces, with limited risks of diffusion of the microorganisms. In any case, the sophisticated microorganisms involved would be poorly adapted to survive in nature. In one scenario, it is not the microorganisms themselves that will be used, but only the complex molecular machinery which will be extracted (or copied) from them.

Nonetheless, the situation – as far as public opinion is concerned – is not entirely favorable. Water is used for drinking, and fears can spring from the possibility to ingest synthetic microorganisms (or parts thereof).

Such technologies might be useful for poor countries due to their low costs of exploitation and the possibilty to develop them on a small scale. Even more so than in other cases, the development of such new organisms will require a sophisticated technology, inaccessible to most countries in which these technologies might be the most useful.

2.5
Soil and Groundwater Decontamination

2.5.1
Introduction

The industrial revolution, which lasted around two centuries in Europe, left its footprints not only in the atmosphere (i.e., CO_2 emission) but also under our feet by creating the so-called brownfields. The common European understanding of the word "brownfield" was defined by former EU networks, such as CABERNET (Concerted Action on Brownfield and Economic Regeneration Network) and CLARINET (Contaminated Land Rehabilitation Network for Environmental Technologies), as: "Brownfields are sites that have been affected by the former uses of the site and surrounding land; are derelict and underused; may have real or perceived contamination problems; are mainly in developed urban areas; and require intervention to bring them back to beneficial use"[2]. The main risk associated with such contaminated areas is the contamination of groundwater, in most cases carrying contaminants further by forming plumes and threatening the water resources.

To date, no common inventory of such contaminated sites exists in the European Union, although some national inventories clearly show the high relevance of this issue on the European level. For example, Germany has about 190 000 sites that are suspected to be contaminated (not including military sites and sites for the production of armaments; Krämer, 2007). In several Member States (e.g., UK) there is an urgent need to slow down the consumption of greenfields and introduce a more sustainable way of brownfield regeneration. This Coordination Action aims at fostering synergies in the various interdisciplinary fields of remediation technology to develop innovative methods to decontaminate brownfields (Sinha *et al.*, 2010). In the Assessment of the European Environment Agency, published in 2007, the EU estimated that there are approximately 250 000 known contaminated sites in Europe that need to be cleaned up. With potentially polluting activities occurring at nearly three million sites, the number of sites requiring remediation will increase by 50% by 2025. The overview of contaminants affecting soil in Europe reveals that more than 60% of the contaminants are organic compounds such as mineral oil hydrocarbons (HC), polyaromatic hydrocarbons (PAH), benzene derivatives (BTEX) and chlorinated hydrocarbons (Figure 2.5). These compounds are more or less susceptible to biodegradation, making bioremediation a key application.

At the same time, many of the organic contaminants show a poor biodegradability due to their diminished bioavailability or their toxic effects on microorganisms.

2) See: http://www.cabernet.org.uk/index.asp?c=1134.

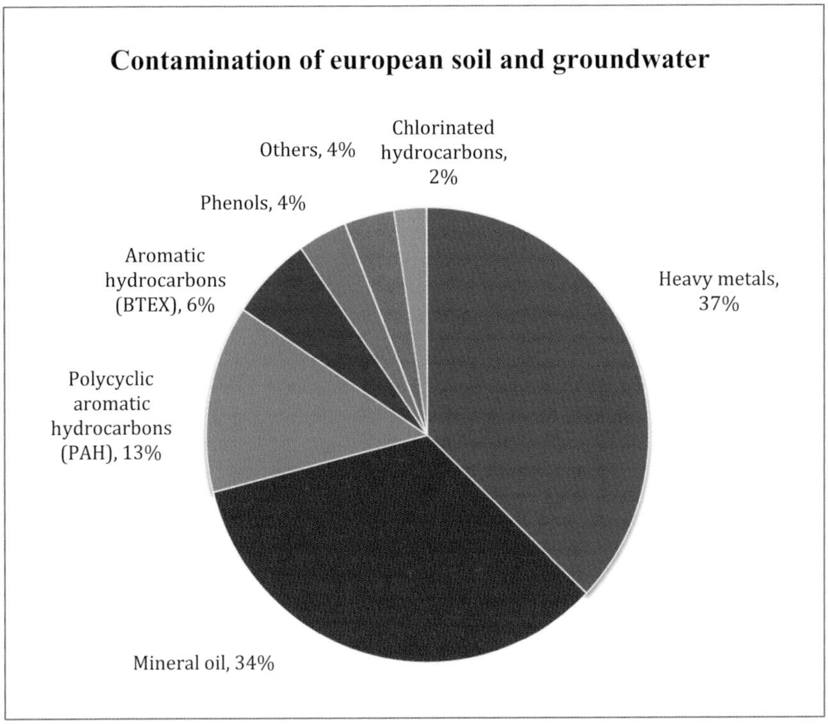

Figure 2.5 Contaminants affecting soil and groundwater in Europe (adapted from EEP, 2007), including data only from Austria, Belgium (Flanders), Czech Republic, Finland, FYR of Macedonia, Greece, Hungary, Italy, Luxembourg, Macedonia, Slovakia, Spain, Sweden.

SB can help to overcome these by creating new organisms which, for example, produce biotensides to enhance the solubility of contaminants, bear new enzymatic activities or which are more resistent to toxic compounds (Thornton *et al.*, 2008).

2.5.2
Economic Potential

As noted above, the contamination of soil with toxic compounds is also a major threat to groundwater and thus to drinking water resources and human beings. Groundwater contamination from a specific site always includes a threat to the neighboring communities. Regeneration of contaminated brownfields will prevent further groundwater contamination. This will help to save costs in the preparation of drinking water and in the dewatering of construction pits (contaminated groundwater pumped from construction pits has to be treated prior to discharge).

In all European countries, there are vast numbers of valuable urban areas that cannot be used due to contaminations. Many factories or gas works originally constructed in the suburban areas of towns in the early twentieth century have gained a central location due to the expansion of towns. In the recent centuries, industry has mostly moved away to commercial and industrial zones in rural areas, leaving behind contaminated but precious estates. New effective techniques for remediation of brownfields will have a considerable economic impact on the remediation industry and on urban communities.

There are many examples, two of which are given in the following:

1) A factory producing office items like pens, ink, markers and so on was located close to downtown Nürnberg. Then the company moved to an industrial zone in the outskirts, leaving behind a contaminated area. The estate was purchased by a group of investors who remediated the estate and built a modern business center.

2) The gasworks near the Olympic Stadium in Munich cover a large area which was heavily contaminated. After it was remediated, the Civic Service of the Munich Municipality built its new office building there and sold many high-value estates to private investors.

Future techniques relying on SB will help to make brownfield remediation no longer a limiting factor compared with greenfield consumption–both from the financial and the decision-making point of view.

The advantages are not limited to abandoned areas. Contaminated industrial areas still in use will also profit from these techniques because remediation is possible under ongoing operation (*in situ*).

2.5.3
Environmental Impact

Remediations have per se a positive effect on the environment because they eliminate or diminish an environmental risk. However, there may be great differences with regard to the technique employed: soil and groundwater remediation with biological methods is considered to be environmental friendly since the elimination of contaminants is carried out under "mild" conditions. Energy is required solely for potential transport and soil movement activities and proesses like aeration/ventilation. The soil retains its natural features and remains intact.

In contrast, thermal soil treatment has a much greater impact on the environment because tremendous amounts of fuel must be consumed additionally to heat soil to the required temperatures of around 400 °C. This is accompanied by large amounts of exhaust gas, which need to be cleaned.

In theory, the cleaned soil is free of contaminants but also free of any form of life and any intact organic matter. Accordingly, any measure including SB that will accelerate biological processes or enable any kind of bioremediation (e.g., for persistent compounds like many polyhalogenated ones) will have a great positive impact on the environment.

Since bioremediation is based on the principles of biodegradation, substances are actually eliminated. This makes the decontamination sustainable. This also holds true for the specific microorganisms used for decontamination: they can reproduce and are therefore renewable and growable on natural substrates.

Certain risks do need to be considered, such as the formation of harmful metabolites during the degradation process. This issue has to be studied carefully when testing new organisms for new degradation pathways. In contrast, SB can also be a useful tool to create tailor-made organisms to degrade certain contaminants in a manner that no harmful metabolites are formed.

Another issue is the release of SB-tailored organisms into the environment. Soil remediation deals with bulk amounts of matter at a scale of hundreds of thousands of tonnes. These masses are used for refilling or construction purposes, after the clean-up goals are reached. Hence, the organisms will be released into the environment. This calls for great caution in excluding any risk associated with the organisms.

2.5.4 Foreseeable Social and Ethical Aspects

The development of efficient techniques for soil and groundwater decontamination will have a dramatic impact, in particular at a time when cultivatable surfaces are declining and land is increasingly slated for the production of biofuels.

The ethical issues concern the risks. Soil and groundwater decontamination occurs in a more or less open space. There is a risk of perturbing the surrounding ecosystem by introducing new organisms. This might fan the fears generated by the introduction into nature of new synthetic living objects, with no possible confinement and boundaries.

Toxic substances will not disappear, but in most cases will be concentrated in wastes. Evacuation and treatment of these wastes will have to be seriously considered.

Treatment of contaminated soil and groundwater using biotools will require less sophisticated machines and protocols than the current technologies. The construction of these biotools, however, will require the competence of specialists. In addition, each decontamination job will have its specificities, which will require preliminary sophisticated studies, with the elaboration of complex models – knowledge and skills that are not easily shared.

2.6 Solid Waste Treatment

2.6.1 Introduction

The average annual amount of domestic waste (industrial waste not included) in Europe is 522 kg/person (Table 2.1). At a population of approximately 500 million,

Table 2.1 Domestic waste in selected EU countries (Eurostat, 2009).

Country	Annual waste (kg/person)	Type of treatment (%)			
		Landfill	Incineration	Recycling	Composting
EU27	522	42	20	22	17
Belgium	492	4	34	39	23
Denmark	801	5	53	24	17
Germany	564	1	35	46	18
UK	572	57	9	22	12
Sweden	518	4	47	37	12
Netherlands	630	3	38	32	28
Italy	550	46	11	11	33
Spain	588	60	10	13	17
France	541	34	36	16	14
Romania	379	99	0	1	0
Bulgaria	468	100	0	0	0

this amounts to about 250 million tonnes of waste every year. Up to 33% or probably more of this amount can be treated biologically (e.g., composting). This is usually done today with the aid of natural organisms like bacteria and fungi generally present on the surface of organic matter.

These processes could be more effective and safer hygienically when SB-based, tailor-made organisms could be used. The benefit would be much greater if the organisms used could transfer organic matter in solid wastes into energy sources such as methane or hydrogen (waste to energy). Such organisms with a very high turnover potential could be designed by SB. It is also conceivable to produce valuable substances like essential amino acids or vitamins (for livestock feed) from organic wastes.

Tailor-made organisms could also help to treat wastes for which no suitable or economic biotechnical methods are available yet. An example for such a waste is scale from steel works, which is usually contaminated with up to 13% oil. In Europe, it amounts to several thousand tonnes per year. Presently, this waste is fed in small doses into furnaces in steel works (otherwise it would cause deflagration in the furnace), creating considerable amounts of off-gas. The energy content of the oil is not used. A pretreatment by microorganisms, which preferably convert the hydrocarbons to methane and/or hydrogen, would create useable energy and a clean valuable waste (iron oxide) which could be fully recycled in steel works. Solid waste conversion to fuels (e.g., methane) clearly carries a great potential for the future.

2.6.2 Economic Potential

Considering the amounts of waste produced every year, effective and more specific methods of biological waste treatment will clearly have a positive economic effect.

A much greater economic benefit is expected when energy or valuable substances are created from organic wastes. If, for example, 10% of the annual amount of domestic wastes in Europe (250 mio tonnes, industrial wastes not considered) could be converted into methane, this would sum up to 25 mio tonnes.

2.6.3 Environmental Impact

SB-based applications are expected to be more specific and more effective. This approach optimizes (fastens) treatment technologies and thus helps to save energy and reduce emissions.

The real benefit for the environment would be the creation of energy sources that would replace fossil energy sources and help to reduce CO_2 emissions. SB applications themselves would not create large amounts of waste. They are sustainable because they are typically based on renewable matter. Special care must be taken when SB-designed microorganisms are used for treatment of waste in open systems like pile, windrows and so on. The environmental risks due to the release of the organisms must be assessed carefully and special measurements must be taken (e.g., treatment in closed vessels) if necessary.

2.6.4 Foreseeable Social and Ethical Aspects

The four established solid waste treatment technologies are: landfill, composting (including windrow composting), incineration and recycling (Giusti, 2009). Additionally, advanced waste treatment technologies are also used, although most of them are still under development. These include: gasification, biodrying, pyrolysis, bioethanol production or waste autoclaving. SB approaches designed to develop GM microorganisms better adapted for solid waste treatment are expected to yield major advances in this field (Asgher *et al.*, 2008). There is clearly a huge range of applications of engineered microorganisms in waste treatment. Examples include modifications allowing adaptation to the high temperatures or the low oxygen contents of composting piles and rows; microorganisms resistant to the wide range of inhibitors that can be found in urban waste; or modified bacteria able to efficiently transform solid waste into biogas or bioethanol. Waste management, along with energy, is one of the key factors of sustainable development. Solid wastes are managed in very different ways, and many Eastern European countries use undesirable strategies such as landfilling to treat virtually all the waste they produce. This situation might be improved by spreading current technologies, mainly

Figure 2.6 Landfill is, unfortunately, the preferred solid treatment technology in many countries, including some European ones. SB applications might facilitate microbial processing and thus promote alternative treatment strategies such as composting or biofuel production from waste (source: Quinte Waste Solutions, 2010).[3]

composting, to these countries. SB approaches could enhance the processing ability of the microorganisms used and thus improve treatment efficiency (Figure 2.6). If waste would be suitably treated, it would results in enormous ecological–and thus societal–benefits (Gentil, Clavreul, and Christensen, 2009). From the point of view of industrial sectors, re-arrangement of workers around the novel technologies is foreseen, but a very limited impact is expected. Although difficult to forecast, it is expected that successful application of SB to solid waste treatment would contribute to a shift in the direction of a truly sustainable development.

Ethical issues. Only very conservative ecologist groups would disagree on using SB microorganisms for waste treatment. But this is certainly going to happen. Note that a false debate between the so-called "3-R" strategy (reduce, reuse, recycle) and SB application might arise: both strategies (creating less waste and treating it more efficiently) have to be considered as complementary. The negative perception of GMOs will also have to be taken into account if outdoor GMO release is expected.

Justice of distribution. Solid waste is associated with industrial production and home consumption, and it is therefore associated with industrial poles and urban areas. Paradoxically, minor waste amounts, such as home waste from villages and rural communities, might be less amenable to SB-based processing because of difficulties involving the availability of the technology. Complex

3) See: www.quinterecycling.org

technologies would have this effect, whereas simple, microbial-based processing would be easier to implement in rural zones.

2.7 CO_2 Recapturing

2.7.1 Introduction

Carbon dioxide has been identified as a prime contributor to global warming, and carbon dioxide recapture is a way of reducing the contribution of fossil fuel emissions to global warming. The recapturing process removes carbon dioxide at the source of its production, as waste from industrial manufacturing processes, and stores it using physical or chemical means. The long-term storage of CO_2 is a relatively new concept, and the first project was the Weyburn project in 2000 (Canadian Geographic, 2008).

Although the term *carbon dioxide capture and storage* has also been used to describe biological techniques such as biochar burial, which uses using adsorbent charcoal in the soil from partly decomposed organic matter to capture CO_2 from the air, it is more frequently used to describe non-biological processes. Currently, the main way of recapturing carbon dioxide is to sink it on a geoengineering scale, under compression into the ground into a variety of underground sites. These include the sea bed, empty aquifers, mines, drill sites, caves, peridotite mantle rock and depleted oil/gas fields. Although this is the dominant method for removing large amounts of carbon dioxide from the atmosphere, there are many infrastructural technical challenges. For example, carbon dioxide purification and compression is important for transport and geological storage in carbon dioxide capture projects. The carbon dioxide purification and compression system must be designed to minimize power consumption while meeting the purity specifications for the carbon dioxide. An indirect reduction in carbon dioxide emissions is also possible by increasing the efficiency of power plants, for example: (i) using low-cost catalysts that help create systems with superior thermodynamics that are not currently practical due to slow kinetics, (ii) employing robust materials that can resist degradation from caustic contaminants in flue gas and (iii) installing advanced capture processes that can dramatically reduce the parasitic energy penalties. Other reductions in carbon dioxide emissions are achieved by power-producing companies through biomass co-firing, where organic matter is added to fossil fuel sources. Since biomass is considered carbon dioxide neutral due to its short time of regeneration compared with fossil fuels, then this (technically) constitutes a renewable source of energy. There are three technological pathways for CO_2 capture, namely post-combustion, pre-combustion capture and oxy-combustion. Carbon dioxide can also be extracted directly from the atmosphere without combustion beforehand (Physorg, 2007). It is important to develop a carbon recapture strategy that is sustainable. The carbon recapture industry is in its earliest stages of development and is predominantly industrially based. The sustainability of these processes as a longer-term approach has not been evaluated.

SB offers an opportunity to develop a carbon recapture industry that works with natural sequestration processes. Synthetic biological systems have the potential to create technologies that can connect directly with and closely mimic the natural biology of our planet. SB-based technologies may also be able to work beyond a biological platform and interface with well developed industrial methods or technologies to generate sustainable outcomes. SB-based technologies may also be able to offer new techniques to capture carbon dioxide based on the engineering of existing biological systems such as artificial leaf technology, designer metabolisms or mass-scale culturing of "designer" algae. Such technologies may also offer new techniques to capture carbon dioxide and transform it into biopolymers, carbohydrates or fuels that can be used to manufacture goods (EPSRC, 2010).

The major challenge for SB techniques in carbon capture is that geoengineering levels of matter are needed to make a difference to carbon dioxide levels. SB simply has not yet been used successfully on this scale. Additionally, the rational engineering of biological organisms on a large scale poses particular challenges related to the general challenges posed by biological systems and supporting infrastructures, but appears to be a promising way of creating biomass from toxic gas emissions. For example, algae-based carbon capture technology (Power Plant CCS, 2010) can remove carbon dioxide from smokestacks and use it along with municipal wastewater to grow algal products that can be later processed into transportation fuels. Engineering challenges in this specific solution include, for example, increasing the surface area for capturing carbon dioxide from smokestacks, bioreactor design to maximize the biofuel yield from algae, SB to enhance the algae's carbon sequestration potential and to generate high-value products efficiently, ecological engineering to design interfaces between the system and externalities, and optimizing the benefit to cost ratio for the entire process. Such an approach does not eliminate the release of carbon dioxide, but it does permit a true biological recycling process. Advances are still needed in the areas of strain development (using traditional and advanced methods), harvesting algae, dewatering and the separation of desired compounds in order for a SB-based approach to be economically competitive.

A less direct way of carbon recapture is by generating alternative fuel sources that are added to fossil fuels as renewable sources of biomass. SB-based manufacturing systems for the combustion of these mixtures will require new infrastructures to support the combustion of biofuel. This is because biomass has a low energy density, is difficult to handle and has a chemical composition that causes significant deposit formation during combustion and post-combustion in combustion engines. This results in acute slagging and fouling of the plant and can lead to reduce heat transfer, blockage within the plant and increased downtime for the removal of deposits.

Several carbon capture and storage projects are currently in development. The first SB-based manufacturing approaches capture carbon dioxide should be commercially viable in the next 10–15 years.

Current carbon capture projects using algae (Power Plant CCS, 2010):

- CEP & PGE, USA, October 2008: One of the most recent algae-inspired projects is being undertaken by Washington-based Columbia Energy Partners LLC (CEP), which hopes to convert carbon dioxide from a coal-fired electricity plant into algal oil.
- Linc Energy & BioCleanCoal, Australia, November 2007: Two Australian firms, Linc Energy and BioCleanCoal, have partnered together in a joint venture to sequester carbon dioxide emissions from Australian coal-fired power stations to use as fuel or fertiliser, even re-burning it to produce additional energy.
- Seambiotic, Israel, August 2007: The Israeli company Seambiotic has found a way to produce biofuel by channeling smokestack carbon dioxide emissions through pools of algae that clean it. The growing algae thrive on the added nutrients and become a useful biofuel.
- Trident Exploration, Canada, August 2007: Trident Exploration Corp. is a natural gas exploration company. The company was examing ways to reduce its CO_2 emissions. Trident approached a number of companies looking for solutions and ended up teaming up with Menova in 2006.
- EniTecnologie, Italy: The objective of the EniTecnologie R&D project on micro-algae biofixation of CO_2 was to evaluate on a pilot scale the feasibility of using fossil CO_2 emitted from a NGCC power plant to produce algal biomass. The biomass would be harvested and then fermented by anaerobic digestion to methane to replace a fraction of the natural gas, with the residual sludge containing most of the nitrogen, phosphorus and other nutrients, recycled back to the cultivation ponds.
- Kolaghat Thermal Power Plant, West Bengal, India, August 2009: An organization based in Kolkata, India, is conducting a pilot project at the Kolaghat thermal power plant and plans to start commercial production of algae bio-fuel by 2010.
- MBD Energy, Australia, August 2009: The Melbourne company MBD Energy is about to introduce technology that allows algae to capture half or more of the greenhouse gases emitted by a power station, at virtually no cost to the utility since it runs only on photosynthesis and also recycles water.
- Arizona Public Service Co., United States, September 2009: Arizona Public Service Co. has landed a $70.5 million US Department of Energy grant to try to feed algae with the carbon dioxide coming from its coal-fired electricity plants.
- RWE, Germany: RWE has studied in detail various options for climate-beneficial recycling and trapping CO_2 in order to identify potentials and obtain recommendations for action. One result of these investigations is the project launched by RWE for binding CO_2 using micro-algae, which are harvested and further explored for conversion into fuel and chemicals.

- E-On Hansa, Germany, November 2007: The German energy group E-On Hansa said it would build a \$3.2 million pilot algae farm for renewable primary products at its Hamburg power plant with support from the city government.
- GreenFuel Technology, United States: Officially reported that it was closing down operations in May 2009. The details provided below are based on the data during the company's operations prior to its closure announcement (www.greenfuelonline.com). Greenfuel began development of their CO_2 capture concept in 2001 with seed money provided through the Massachusetts Institute of Technology (MIT) entrepreneurship competition.
- NRG Energy, United States, April, 2007: NRG Energy and GreenFuel Technologies started testing GreenFuel's algae to biofuels technology at a 1489-megawatt coal power plant in Louisiana. GreenFuel's Emissions to Biofuels™ process uses engineered algae to capture and reduce flue gas carbon dioxide (CO_2) emissions into the atmosphere. The algae can be harvested daily and converted into a broad range of biofuels or high-value animal feed supplements, according to the company.
- ConocoPhillips: Involved in a \$5 million, multi-year sponsored research agreement with the Colorado Center for Biorefining and Biofuels, a research center of the Colorado Renewable Energy Collaboratory. The first project involves converting algae into renewable fuel. Algae can potentially offer substantially higher yields than other oil-producing crops. Moreover, because algae are not food crops, there is less competition for them in the food chain and less risk of adverse affects on food prices. Similarly, algae do not compete for productive farmland. They can grow in freshwater, saline water or even sewage. Finally, algae can utilize CO_2 emissions from industrial operations, making algae an interesting potential carbon capture mechanism. It will probably take another 10–15 years before an algae to fuel process becomes commercially viable (ConocoPhillips, 2008).

2.7.2 Economic Potential

Carbon sequestration has long been regarded as one possible solution to help mitigate the effects of global warming; more recently, new technological developments in this field, especially sustainable ones, are designed to kick-start the global economies out of the current economic recession (BBC News, 2005).

There is great optimism for sustainable technologies such as SB-based applications to produce more environmentally friendly ways of manufacturing and generating alternative fuel sources to biofuels. Accordingly, this research and development area is heavily funded by national governments, despite the industry having no well documented commercial track record.

Since there is no established industry in this field, the economic model in which the SB-based technologies will be competing is the carbon economy. This complex system is influenced by a number of factors and operates on the economic principles of carbon trading.

Carbon trading is a market mechanism intended to tackle global warming. Though it dates back to 1989, it only took off as a market after the Kyoto Protocol was signed (BBC News, 2006). Under the Kyoto treaty, which came into force in February 2005, industrialized countries must reduce total greenhouse gas emissions by an average 5.2% compared with 1990 levels between 2008 and 2012. The most important greenhouse gas contributing to global warming is carbon dioxide, which is mainly emitted by burning fossil fuels. Under Kyoto, each participating government has its own national target for reducing carbon dioxide emissions. The key idea behind carbon trading is that the sources of carbon dioxide are far less important than total amounts. Therefore, instead of the rigidly forcing reduced emissions country by country (or company by company), there is a choice provided by the market economy: either spend the money on cutting pollution (emissions), or else continue polluting (emitting) but use the money to pay for someone else to cut their pollution. In theory this enables emissions to be cut with a minimum price tag. Exact figures are hard to come by because the market is still fairly new, because data is not easily available and because several different schemes exist, not all directly comparable. The World Bank, one of the main players in carbon financing, estimates the value of carbon traded in 2005 to be about \$10 billion. The Bank believes the carbon market has the potential to bring more than \$25 billion (£14 billion) in new financing for sustainable development to the poorest countries and the developing world. Trading firms, brokers and banks are among those expected to make money through commissions for organizing carbon deals. The Bank's own carbon finance fund has more than doubled from \$415 mio in 2004 to \$915 mio last year.

2.7.2.1 How Is Carbon Traded?

There are two main trade models for carbon emissions in the future commodity market. The first is known as a cap and trade scheme: the emission limit is set and the quota can then be traded. Under the Kyoto Protocol, developed countries can trade between each other. One carbon credit is equal to one tonne of carbon dioxide or, in some markets, carbon dioxide-equivalent gases. Carbon trading is an application of an emissions trading approach. Greenhouse gas (GHG) emissions are capped and then marketed to allocate the emissions among the group of regulated sources. This scheme aims to drive industrial and commercial processes in the direction of low emissions or less carbon-intensive approaches. In addition, credits will be generated from GHG mitigation projects which in turn can be used to finance other carbon reduction schemes worldwide. The European Trading Scheme (ETS) is a cap and trade scheme and the largest company-based scheme. It is mandatory and includes 12 000 sites across the 25 European Union member states. It came into force in 2005 and covers heavy industry and power generation, including non-European companies. There are also voluntary cap and trade schemes. The Chicago Climate Exchange (CCX) is such a scheme. Interest in carbon trading at the regional level is increasing in America, even though the United States government has decided not to ratify Kyoto. The United Kingdom also has its own voluntary scheme, for which companies cut their emissions in return for incentive payments.

The second main model of trading carbon is by crediting projects that compensate for or "offset" emissions. The Kyoto Protocol's clean development mechanism (CDM), for example, allows developed countries to gain emissions credits for financing projects based in developing countries. A Kyoto mechanism called joint implementation (JI) also involves project-based schemes, whereby one country can receive emissions credits for financing projects that reduce emissions in another developed country. Compliance is critical for this model. Obligated by the Kyoto Protocol, the developed countries have 100 days after final annual assessments to pay for any shortfall – by buying credits or more allowances via emissions trading, while penalties will be applied If failure to do so. However, in voluntary schemes, by contrast, this is not the case. Additional economic benefits from "carbon trading" include the actual purchase of fuel and running costs associated with the drive towards more efficient technologies and efficient use of technologies.

There is also a market in the sale of carbon dioxide to the oil industry as part of an enhanced oil recovery scheme. Here, carbon dioxide is pumped into oil fields under pressure to extract difficult to reach fossil fuel supplies (Scientific American, 2009). According to 2009 figures produced by the Scottish government, CO_2 is a cost to enhanced oil projects in the range of £20–40 ($28–56) per tonne (Scottish Government, 2009).

There is a huge global demand for effective technologies that can capture carbon dioxide emissions as part of a sustainable global economy. Weaknesses are that the current CO_2 capture technology is expensive, in terms of both capital and operating cost. This puts a heavy reliance on research and development efforts to improve the economic profile of the technologies. Also, the existing technology is very new and therefore the value of investment is uncertain. There is also much skepticism about the potential of carbon recapture processes to make any meaningful difference to the atmospheric levels of the gas, whether at the industrial scale or using SB manufacturing processes. The competition for research and development grants is fierce. Additionally, complex market forces working on an emerging industry make commercial investment very risky. The global political climate and trade agreements are crucial to keep an effective carbon trading system operational because this relies on the compliance of its members. Opportunities in this market will reside in solutions that appear attractive to investors, even if they are not tried and tested. As this market is not heavily regulated, this is a trade-off in terms of investment risk. Threats to the market are based in cynicism and skepticism for the "reality" of global warming, combined with the newness of the developing technologies and the variability of products. Also, the capacity for industray to meet a sudden demand for the technology is not established, with no established product or technological lifecycle information to guide manufacturers or investors.

2.7.3 Environmental Impact

SB-based manufacturing approaches to carbon capture create more biological-like than industrial processes. These processes are more sustainable as they are compatible with natural systems. SB offers the potential to create a variety of different "species" of solutions to carbon-fixing approaches. This helps avoid a "monoculture" of technologies, which may result in the accumulation of toxins or overuse of vital resources such as water to produce the end result. SB-based approaches could produce novel solutions that have the potential to evolve and adapt to changing environments as well as vary with geographical locations and environmental conditions. Such applications may provide direct carbon dioxide recapture and fixation into biological materials that can be recycled, used as construction materials or consumed in the ecosphere and therefore reduce industrial waste. These applications may help augment the performance or quality of existing alternative energy sources and generate more efficient and environmentally friendly biofuels and novel approaches. Examples include hybrid chemical and biological-like processes using bottom-up approaches with protocells. However, the technologies to support the development of these SB-based systems require a heavy investment of energy and resources. This is because bulk manufacturing of these agents is needed to provide scales that are capable of geoengineering-scale results. Moreover, they consume natural resources, so that their local and global ecological impact has to be carefully considered.

The sustainability of an agent implies a lifecycle that enables it to thrive while not affecting successive generations of organisms. No doubt, nutrients will need to be regularly added to a manufacturing process, and growth substrates will need to be refreshed. These external agents will need to come from sustainable sources themselves, for example, the local water supply. Since SB-based manufacturing systems are still in the research and development stage, it is not possible to fully evaluate their sustainability or the extent to which their health and functioning would be affected while being used in a manufacturing context.

The whole carbon cycle of these new SB-based manufacturing systems must be taken into account. Use of land and water supplies needs to be considered carefully, along with the lifecycle of the organisms used for carbon fixation. The newly manufactured organisms are likely to be fragile. This calls for examining their biological design and context in determining how sustainable they are. If they are in containers, then the organisms may be susceptible to inhibitors. If they are in the environment, they may be susceptible to predation or be out-competed in natural systems.

Modified organisms bear the risk that the technology will have a competitive impact on the local environment and natural ecosystems. Since the organisms will be needed in large quantities, the risk of leakage or a spill in these manufacturing plants is significant. Established procedures will need to be in place to detoxify land that has been exposed to a spill. The technology exhibits a degree of unpredictability because it involves a living system which can unexpectedly and rapidly mutate or change its function. Batches will need to be tested and monitored for their consistency and homogeneity. It is also possible that the technology will promote climate change as a result of deforestation and establishment of monocultures; this is a consequence of intensive biomass farming that drains water and land for the production of food.

Living organisms are major players of carbon dioxide fixation. Beside those found in plants and algae, several other pathways have been discovered that permit carbon dioxide fixation using means other than standard photosynthesis. It seems difficult to see how SB-based approaches could out-compete plants and algae at the moment. Nonetheless, improving metabolic pathways by SB-based approaches could lead to carbon dioxide recapture within the internal metabolism of synthetic cell factories. A systematic implementation would considerably improve the carbon burden of SB factories. It is currently unclear whether this will be feasible. However, SB-based approaches are leading to novel mechanisms for carbon sequestration and recycling when linked to inorganic chemistry. A case in point is the work being conducted by the Cronin group on the Solar Fuels program at the University of Glasgow. New materials are being developed for the production of biofuels, for example, the research led by Dr Frank Marken at the Universities of Bath, Bristol and the West of England (EPSRC, 2010). Such new processes may play a significant role in increasing the efficiency of carbon fixation and its metabolism into reusable organic compounds. Research and development in this field is in its earliest stages and requires support to fully understand the true impact of these novel and increased efficiency approaches.

2.7.4 Foreseeable Social and Ethical Aspects

A social understanding and engagement with SB manufacturing techniques requires cooperation and understanding between people from different cultures and backgrounds. Such an understanding is an aspect of being globalized and being able to communicate the benefits of national investments to international audiences. This calls for engagement with humanistic outcomes for the long-term survivability of the planet and a better quality of life worldwide. Raising global public awareness and becoming engaged in a progressive conversation about the environment are essential steps. This is because a carbon economy requires society to engage with the concept of climate change. Society is also called upon to believe that it is both ethical and logical to try to change the current situation through technological intervention, especially when new (SB) techniques are employed. Educational and community-based systems are necessary to ensure that these rational and mediated conversations take place in collaboration with those technologists and scientists developing the new approaches.

Sustainable development is a humanitarian concept aimed at improving the quality of life and the welfare of humanity equitably. This approach must meet the needs of future generations under the conditions of limited world resources and under the far-reaching effects of current industrial activity. Implementation of new technologies will require fundamental changes in the way people go about their daily lives and what kinds of jobs they will have. It will also depend on the underlying economic structure production and consumption profiles, technologies, institutions and organization of the manufacturing systems that use the technology. The role of changes in technology is particularly crucial and will serve as a key driver of the whole transformation of developmental models across society in general.

A couple of issues are related to carbon trading and the carbon economy. Carbon trading, whether between companies or countries, only works if everyone agrees to participate in the project and if emissions are reduced enough to contain global warming. Creating a market does not, by itself, reduce emissions. Moreover, the benefits could be severely limited if trading is not comprehensive. The United States, the world's largest energy comsumer (per capita), excluded itself by choosing not to ratify Kyoto. The United States is also the biggest emitter of carbon emissions (per capita) today. At the same time, China, which is projected to exceed the United States in emissions by 2050, has no obligation to reduce emissions.

There is an uneven economic playing field in a carbon trading system. Even within trading schemes such as the ETS, the emissions of whole sectors' are excluded, such as transport, homes and the public sector. Aviation is the fastest-growing source of CO_2 emissions, and some experts have calculated that if it were included, the United Kingdom's entire allowance would soon be used up. For a carbon trading system to work, it would have to become much broader, perhaps even embracing personal carbon allowances for individuals.

Critics say trading carbon condones the idea of "business as usual" and fails to emphasize the need to invest in renewable energies and move away from fossil fuels.

Is it ethical for national and international governments to heavily support an unproven technology for applications that ultimately are designed to have a significant environmental impact? Who is making the decisions? Who can play God with the environment and corrupt the "sacred" natural world? Environmentalists argue that the practice of carbon trading trivializes the scale of the underlying problem. Also, carbon dioxide represents only part, though more than 70%, of all greenhouse gases. An increasing number of scientists are saying that the carbon dioxide ceilings under the treaty are too high, perhaps far too high, to help avert serious climate change.

There will probably be widespread inequalities in access based on economics, manufacturing infrastructure and geography. The developed world will benefit and the developing world will lose out, particularly in the competition for land and water used to produce the biomass that underpins a SB manufacturing industry. Specific objections by the developing countries are that the carbon trading system merely allows polluters to buy their way out of reducing their emissions. This multibillion-dollar pollution trading mechanism is claimed to be privatizing the air and commodifying the atmosphere. Groups such as the International Indigenous Peoples' Forum on Climate Change oppose the carbon market and condemn it as "a new form of colonialism." The "reducing emissions from deforestation and degradation" may result in more violations of indigenous peoples' rights as carbon traders take control of natural resources such as forests in order to purchase land for the development of SB-based manufacturing plants with monoculture tree plantations that cause environmental and social devastation.

2.8 Recommendations for Bioremediation

Bioremediation is an area with a great potential for benefits provided by SB. Bioremediation is usually applied on materials with a massive occurrence such as solid (organic) wastes, sewage, industrial wastewater, contaminated soil or contaminated groundwater, all measured in millions of tonnes or cubic meters. We believe that SB has the potential to create tools to improve the treatment methods, saving costs and environmental resources. Moreover, it can provide methods to produce energy or valuable goods from waste or wastewater. It can also provide tools for making fresh- or drinking water either from contaminated water or seawater. Another possible field of application is the production of biosensors to monitor environmental goods and hazards. Based on a differentiated evaluation, we conclude that biosensors provided by SB tools would have a great positive effect on the environment because they will help to survey environmental hazards more precisely and effectively. Their economic and social impact, however, is rather low because they are niche products.

SB-based approaches may provide a way of capturing, storing and recycling carbon dioxide. This may be through the re-engineering of existing organisms or the creation of novel carbon processes. Great importance is attached to bottom-up approaches in which inorganic chemistry is linked to living processes through agents such as the emerging protocell technology.

SB-based carbon capture may not be able to lower carbon dioxide to a level that completely remediates the current escalating levels that are being released through fossil fuel consumption. This is because geoengineering-scale approaches are necessary. The approach does, however, offer the possibility of carbon capture and recycling, a task which current industrial-scale processes cannot do.

One key recommendation is to support SB approaches because of the scale of the problem with carbon dioxide emissions and the urgent need for remediation. This will yield the next generations of carbon capture technologies, technologies that will fo beyond storing the carbon dioxide to recycling it into fuels and biopolymers with a positive environmental impact.

Another positive impact, particularly on the environment, can be expected for soil and groundwater remediation, especially with regard to enhancing the clean-up efficiency and developing new methods. In this field, the economic and social impacts are rather moderate because it is a specific field with a limited scope of time. We expect the strongest impact for solid waste and wastewater treatment and for water desalination. The importance of the latter cannot be overstated in a world where billions of people have no access to clean drinking water or freshwater for agricultural use. Solid waste and wastewater treatment also bear a great potential for improvement by SB due to their sheer volume and considerable organic content. We therefore strongly recommend to support the development of these three issues. One potential constraint should be mentioned: solid waste and wastewater cannot be treated in sealed vessels or rooms, simply due to the enormous volumes involved. They have to be treated openly in piles or basins. Therefore, the use of engineered cells may create a problem of interaction with the environment, an important point to keep in mind. At the same time, we expect no limitations for the use of non-proliferative systems such as enzymes or protocells created using SB.

References

2. BIOREMEDIATION

2.1 BIOREMEDIATION IN GENERAL

BAFU (2009) Micropollutants in the aquatic environment, BAFU (Swiss Environmental Agency).

Cao, B., Nagarajan, K., and Loh, K.C. (2009) Biodegradation of aromatic compounds: current status and opportunities for biomolecular approaches. *Appl. Microbiol. Biotechnol.*, **85**, 207–228.

Kavamura, V.N., and Esposito, E. (2010) Biotechnological strategies applied to the decontamination of soils polluted with heavy metals. *Biotechnol. Adv.*, **28**, 61–69.

Rojo, F. (2009) Degradation of alkanes by bacteria. *Environ. Microbiol.*, **11**, 2477–2490.

Shimao, M. (2001) Biodegradation of plastics. *Curr. Opin. Biotechnol.*, **12**, 242–247.

2.2 DETECTION OF ENVIRONMENTAL POLLUTANTS (BIOSENSORS)

ACT (2002) ACT workshop: biosensor for harmful algal blooms, http://www.act-us.info/Download/Workshop_Reports/ACT_WR02-01_HAB.pdf (accessed 28 May 2010).

Beep-c-en (2010) Biosensor for effective environmental protection and commercialization: enhanced, http://www.beep-c-en.com/project.php (accessed 28 May 2010).

BIO4EU (2007) Consequence, opportunities and challenges of modern biotechnology for Europe (BIO4EU) Task 2: case studies, report the impact of industrial biotechnology applications, http://bio4eu.jrc.ec.europa.eu/documents/Bio4EU-Task2Annexindustrialproduction.pdf (accessed 28 May 2010).

Biosensor at EPA (2009) Biosensors for environmental monitoring, http://www.epa.gov/heasd/edrb/biochem/env-monit.html (accessed 28 May 2010).

Ehrentreich-Förster, E., Orgal, D., Krause-Griep, A., Cech, B., Erdmann, V.A., Bier, F., Scheller, F.W., and Rimmele, M. (2008) Biosensor based on-site explosives detection using aptamers as recognition elements. *Anal. Bioanal. Chem.*, **391**, 1793–1800.

Fuji-Keizai Report (2004) Biosensor market, R&D commercial implication, http://www.researchandmarkets.com/reportinfo.asp?report_id=63374&t=d&cat_id (accessed 28 May 2010).

Harms, H., Wells, M.C., and van der Meer, J.R. (2006) Whole-cell living biosensors: are they ready for environmental application? *Appl. Microbiol. Biotechnol.*, **70** (3), 273–280.

Khalil, A.S., and Collins, J.J. (2010) Synthetic biology: applications come of age. *Nat. Rev. Genet.*, **11**, 367–379.

Li, J., Wang, Z., Ma, M., and Peng, X. (2010) Analysis of environmental endocrine disrupting activities using recombinant yeast assay in wastewater treatment plant.

Bull. Environ. Contam. Toxicol., **84** (5), 529–535.

Looger, L.L., Dwyer, M.A., Smith, J.J., and Hellinga, H.W. (2003) Computational design of receptor and sensor proteins with novel functions. *Nature*, **423** (6936), 185–190.

Luong, J.H., Male, K.B., and Glennon, J.D. (2008) Biosensor technology: technology push versus market pull. *Biotechnol. Adv.*, **26** (5), 492–500.

Nakamura, H., Shimomura-Shimizu, M., and Karube, I. (2008) Development of microbial sensors and their application. *Adv. Biochem. Eng. Biotechnol.*, **109**, 351–394.

Sinha, J., Reyes, S.J., and Gallivan, J.P. (2010) Reprogramming bacteria to seek and destroy an herbicide. *Nat. Chem. Biol.*, **6**, 464–470.

Team Edinburgh IGEM (2006) Arsenic biosensor, http://parts.mit.edu/wiki/index.php/Arsenic_Biosensor (accessed 28 May 2010).

Tescione, L., and Belfort, G. (1993) Construction and evaluation of a metal ion biosensor. *Biotechnol. Bioeng.*, **42** (8), 945–952.

Win, M.N., Liang, J.C., and Smolke, C.D. (2009) Frameworks for programming biological function through RNA parts and devices. *Chem. Biol.*, **16** (3), 298–310.

2.3. WATER TREATMENT

Shannon, M., Bohn, P.W., Elimelech, M., Georgiadis, J.G., Mariñas, B.J., and Mayes, A.M. (2008) Science and technology for water purification in the coming decades. *Nature*, **452**, 301–310.

Tal, A. (2006) Seeking sustainability: Isrel's evolving water management strategy. *Science*, **313**, 1081–1084.

2.4. WATER DESALINATION WITH BIO-MEMBRANES

Cau, X., Huang, X., Liang, P., Xiao, K., Zhou, Y., Zhang, X., and Logan, B.E. (2009) A new method for water desalination using microbial desalination cells. *Environ. Sci. Technol.*, **43**, 7148–7152.

Shannon, M., Bohn, P.W., Elimelech, M., Georgiadis, J.G., Mariñas, B.J., and Mayes, A.M. (2008) Science and technology for water purification in the coming decades. *Nature*, **452**, 301–310.

Tal, A. (2006) Seeking sustainability: Israel's evolving water management strategy. *Science*, **313**, 1081–1084.

2.5. SOIL DECONTAMINATION

EEP (2007) Overview of contaminants affecting soil and groundwater in Europe, http://www.eea.europa.eu/data-and-maps/figures/overview-of-contaminants-affecting-soil-and-groundwater-in-europe (accessed 28 May 2010).

Krämer, L. (2007) *EC Environmental Law*, 6th edn, Sweet & Maxwell, London.

Sinha, J., Reyes, S.J., and Gallivan, J.P. (2010) Reprogramming bacteria to seek and destroy an herbicide. *Nat. Chem. Biol.*, **6**, 464–470.

Thornton, I., Farago, M.E., Thums, C.R., Parrish, R.R., McGill, R.A., Breward, N., Fortey, N.J., Simpson, P., Young, S.D., Tye, A.M., Crout, N.M., Hough, R.L., and Watt, J. (2008) Urban geochemistry: research strategies to assist risk assessment and remediation of brownfield sites in urban areas. *Environ. Geochem. Health*, **30** (6), 565–576.

2.6. SOLID WASTE TREATMENT

Asgher, M., Bhatti, H.N., Ashraf, M., and Legge, R.L. (2008) Recent developments in biodegradation of industrial pollutants by white rot fungi and their enzyme system. *Biodegradation*, **19** (6), 771–783.

Eurostat (2009) Half a ton of municipal waste generated per person in the EU27 in 2007, http://epp.eurostat.ec.europa.eu/cache/ITY_PUBLIC/8-09032009-BP/EN/8-09032009-BP-EN.PDF (accessed 28 May 2010).

Gentil, E., Clavreul, J., and Christensen, T.H. (2009) Global warming factor of municipal solid waste management in Europe. *Waste Manag. Res.*, **27** (9), 850–860.

Giusti, L. (2009) A review of waste management practices and their impact on human health. *Waste Manag.*, **29** (8), 2227–2239.

Quinte Waste Solutions (2010) www.quinterecycling.org (accessed 28 May 2010).

2.7. CO_2 RECAPTURING

BBC News (2005) Funds for greenhouse gas storage, http://news.bbc.co.uk/2/hi/science/nature/4089538.stm (accessed 28 May 2010).
BBC News (2006) Q&A the carbon trade, http://news.bbc.co.uk/2/hi/business/4919848.stm (accessed 28 May 2010).
Canadian Geographic (2008) Burying the problem, http://www.canadiangeographic.ca/magazine/JF08/indepth/weyburn.asp (accessed 28 May 2010).
ConocoPhillips (2008) Sustainable development: renewable fuels, http://www.conocophillips.com/EN/susdev/environment/cleanenergy/alternativeenergy/Pages/RenewableFuels.aspx (accessed 28 May 2010).
EPSRC (2010) From pollutant to profit: nanoscience turns carbon on its head, http://www.epsrc.ac.uk/newsevents/news/2010/Pages/nanoscienceturnscarbononitshead.aspx (accessed 28 May 2010).
Physorg (2007) First successful demonstration of carbon dioxide air capture technology achieved, http://www.physorg.com/news96732819.html (accessed 28 May 2010).
Power plant CCS (2010) Algae based carbon capture, http://www.powerplantccs.com/ccs/cap/fut/alg/alg.html (accessed May 2010).
Scientific American (2009) Enhanced oil recovery: how to make money from carbon capture and storage today, http://www.scientificamerican.com/article.cfm?id=enhanced-oil-recovery (accessed 28 May 2010).
Scottish Government (2009) CO_2 enhanced oil recovery, http://www.scotland.gov.uk/Publications/2009/04/28114540/ (accessed 28 May 2010).

Further Reading

Kotrba, P., Najmanova, J., Macek, T., Ruml, T., and Mackova, M. (2009) Genetically modified plants in phytoremediation of heavy metal and metalloid soil and sediment pollution. *Biotechnol. Adv.*, **27**, 799–810.
Sandia Corporation (2010) Implementation of the National Desalination and Water Purification Technology Roadmap, http://www.sandia.gov/water/desal/tech-road/index.html (accessed 28 May 2010).

3
Biomaterials

Lei Pei, Rachel Armstrong, Antoine Danchin, and Manuel Porcar

3.1
Biomaterials in General

3.1.1
Introduction

A biomaterial is a natural or man-made material which forms the whole or part of a living structure or biomedical device with a natural function (Enderle, Blanchard, and Bronzino, 1999). Biomaterials have a wide range of different functions and applications and include biomass-derived basic chemicals, fine chemicals, bulk chemicals and bioplastics/polymers.

Biomaterials have applications in environmental contexts, for example, sewage slurry, and as alternative building materials. An adhesive named PureBond, which contains soya proteins modified to resemble the adhesive protein mussels use to attach to rocks, is now used in 40% of the plywood and composite wood produced in the United States, according to its manufacturer, Columbia Forest Products of Greensboro, North Carolina (Columbia Forest Products, 2010). SB has the potential to greatly stimulate the biomaterials market across a wide range of applications in major commercial fields and specific interest areas of energy, environment and agriculture to radically stimulate industry growth. SB-based manufacturing processes open up the possibility of manufacturing high-quality biomaterials on a larger scale of production using techniques such as mass cell and the potential to engineer biomaterials that are nano bio info cogno (NBIC) convergent. Such biologically compatible information processing systems enable biomaterials to perform as microprocessors and operate a range of control functions. This could involve signaling and actuators, which replace or modify their natural equivalents. A significant challenge to this industry relates to infrastructural challenges to develop new SB-based engineering approaches. Such new biomaterials are currently being used in the built environment. A collaboration between the global architectural design firm HOK and the biologically motivated innovation company Biomimicry Guild is expanding the mainstream application of bio-inspired design into the planning and design of architecture, communities and cities globally

Synthetic Biology: Industrial and Environmental Applications, First Edition. Edited by Markus Schmidt.

(Hills, 2008). The intention is to create more sustainable and energy efficient buildings.

SB-based approaches can offer more efficient bio-compatible materials or create new ones that are biodegradable. This will help increase the range of approaches by generating more environmentally friendly manufacturing practices.

The USDoE (2004) issued a report that identified 12 building-block chemicals that can be produced from sugars via biological or chemical conversions. These building blocks can be subsequently converted to numerous high-value bio-based chemicals or materials. Building-block chemicals are molecules with multiple functional groups that have the potential to be transformed into new families of useful molecules. The 12 sugar-based building blocks are:

1) Four-carbon 1,4-diacids (succinic, fumaric, malic);
2) 2,5-Furan dicarboxylic acid (FDCA);
3) 3-Hydroxy propionic acid (3-HPA);
4) Aspartic acid;
5) Glucaric acid;
6) Glutamic acid;
7) Itaconic acid;
8) Levulinic acid;
9) 3-Hydroxybutyrolactone;
10) Glycerol;
11) Sorbitol (alcohol sugar of glucose);
12) Xylitol/arabinitol (sugar alcohols from xylose and arabinose).

The chemical building blocks play an important intermediate role in the production of useful chemicals from primary biomass. The first part is the transformation of sugars or other raw material to the building blocks. The second part is the conversion of the building blocks to secondary chemicals or families of derivatives. The advantage lies in the fact that, with these building blocks, SB needs only a limited number of metabolic modules to generate them and another set of metabolic modules to produce the end product. This eliminates the need to construct a new full metabolic pathway for each useful chemical. Also, the 12 building blocks can serve as bio-commodities for future bio-trade. See Figure 3.1 and Table 3.1 for more details.

3.1.2
Economic Potential

The field has experienced a steady and strong growth over its history, with many companies investing large amounts of money into the development of new products. Biomaterials have experienced their greatest impact in medicine (Markets&Markets Report, 2009). Biomaterials products had a market value of US$ 25.5 billion in 2008, and the biomaterial device market at US$ 115.4 billion in the same year. The future market value is expected to reach US$ 252.7 billion in 2014. This massive revenue potential indicates the immense opportunity

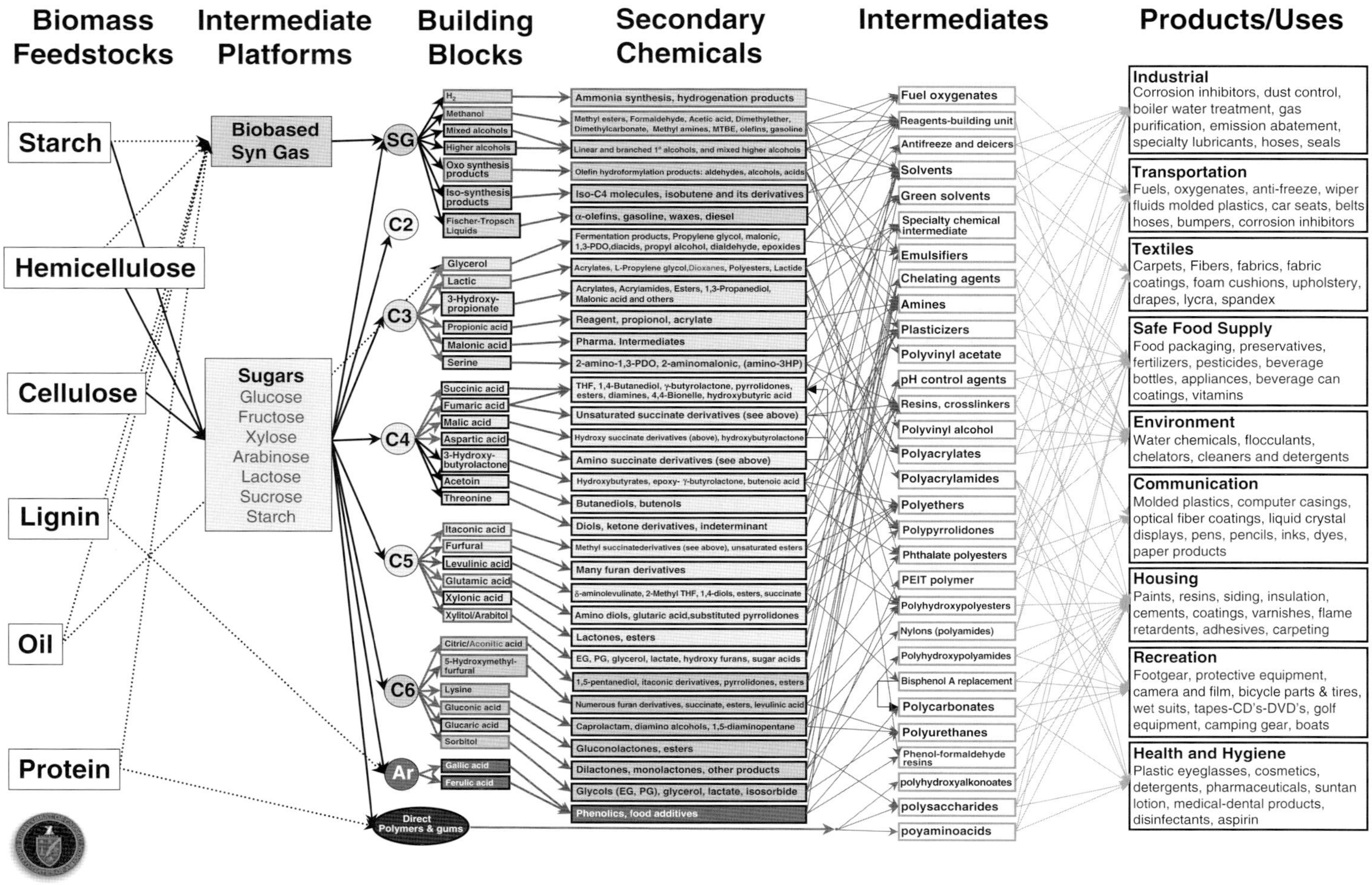

Figure 3.1 Schematic view of bio-based product flow chart from biomass feedstocks to building blocks and then to useful products (Source: USDOE 2004).

Table 3.1 Bulk chemical production in commercial scale.

Monomer	Feedstock	Scale	Company	Application
Methacrylic monomers	Bioethonal	NA	Rohm & Haas[a)]	PMMA
Acrylic acid	Sugar	NA	Cargill[b)]	Fibers
Ethylene	Sugarcane	200 000 ton	Braskem[c)]	Bio-PVC
Ethylene	Bioethonal	350 000 ton	Dow	Bio-PVC
Ethylene	Ethonal	60 000 ton	Solvay[d)]	Bio-PVC
1,3-propanediol (PDO)	Corn	45 000 ton	DuPont and Tate & Lyle[e)]	Bio-polyester
Polyol	Vegetable oil	1 Mio. ton	Bayer[f)]	Intermediate of polyurethanes (PU)
Succinic acid	Glucose	2 000 ton	Bioamber (DNP)[g)]	

a) See: http://greentechnolog.com/2007/04/bioprocess_for_monomer_from_cellulosic_materials.html
b) See: http://www.novozymes.com/en/MainStructure/PressAndPublications/PressRelease/2008/Cargill+cooperation.htm
c) See: http://www.plasticstoday.com/articles/braskem-details-sugar-cane-pe-plans
d) See: http://www.solvinpvc.com/static/wma/pdf/1/2/1/0/0/Press_release_Brasilian_SolVinPVC_EN_141207.pdf
e) See: http://www2.dupont.com/Automotive/en_US/news_events/article20060619b.html
f) See: http://www.research.bayer.com/edition-20/20_renewable_raw_materials.pdfx
g) See: http://www.bio-amber.com/succinic_acid.html

and attracts more investment into the field. It is predicted that the medical biomaterials market is to grow at a combined annual growth rate (CAGR) of 15% and the value of the bioplastics market was estimated at US$ 250 billion (London Metal Exchange, 2010). Extrapolating from the rapid growth area in medicine, expectations for market growth across all areas of biomaterials applications are high. SB is regarded as having the potential to create new industries that will turn around the world economy. Improved environmental benefits form the most important factor stimulating market growth for biomaterials. Improvement in manufacturing technologies and new products at competitive prices will be the major driving force to future market growth. There is a pressing need for the development of appropriate regulatory frameworks for biomaterials that support commercial development and market growth.

3.1.3 Environmental Impact

Many biomaterials are currently in use: wood, cotton, linen, hemp, wool, horn, leather and so on. Paper is essentially based on biomaterials. Polyhydroxybutyrate is already being used to make bags. Cellulose is one biomaterial of interest that might be produced by SB techniques. It could be engineered to create paper by producing a biofilm. Silk-based proteins (Widmaier *et al.*, 2009), which have a significant range of interesting applications, are another promising biomaterial. All these materials are biodegradable (slowly, though, as by definition a material needs to be, at least for some time, stable). One idea would be to add value to certain types of waste to transform them into biomaterials (this has already been developed in some situations, especially with protein-rich material). Part of the impact will depend on the chemicals (especially minerals) included in the final product (e.g., the role of chromium in leather tanning). The environmental impact of producing these materials may be less than producing them in traditional agriculture and/or through chemical processes.

The industrial production of biomaterials could have environmentally friendly effects: less air and water pollution, less energy consumption, fewer raw materials needed, and/or the related products could be eco-friendly because of their biodegradability. For those biomaterials made from biopolymers, the environmental impact of biomaterials will be illustrated in detail in the following section.

As a general principle, the environmental impact of biomaterials will be evaluated by adjusted life cycle assessments with a focus on:

1) Raw material supply, including biomass production and land-use;
2) Energy requirements for industrial processing;
3) Durability of these naturally derived materials compared to those derived from fossil sources;
4) End of life issues, including recyclability, composting, recovery and re-use.

3.1.4
Foreseeable Social and Ethical Aspects

The perspective to replace existing chemistry by "green chemistry" (Carothers, Goler, and Keasling, 2009; Keasling, 2007) will be positively received by the public, especially considering the negative image of chemistry – too often for irrational reasons – in society. In addition, "green chemistry" has always occurred in nature (e.g., oil production in earth history), and humans have mastered it to produce cheese, wine, beer, bread and so on.

Green chemistry will not spell the end of chemical industrial complexes is no doubt an illusion. Even if the size of industrial plants is reduced by using microorganisms, the necessity for sophisticated and huge devices to purify the active substances will remain. The danger in manipulating chemical substances will not fully disappear: the quantities of intermediate reactants will be reduced, but if the final product is dangerous *per se*, the problem will remain unchanged.

Developing countries may suffer rather than benefit from these developments. They may struggle to complete with SB-based processes developed by industrial countries. The microorganisms themselves might easily circulate, but it is difficult to share the sophisticated technologies required for the interface between these microorganisms and the more traditional chemical part, the purification procedures and the biological state of the microorganisms.

Biosafety issues with biomaterials are also important with respect to the longevity of the materials within the environment and whether they have any local or systemic toxic effects.

Widespread inequalities in access are expected due to economics, existing health care systems and geography. This refers both to the provision of the application and the nature of biomaterials for medical applications. Since some of the biomaterials are essentially "living", they will need an ongoing care and assessment infrastructure to ensure their optimum functioning and longevity. Inequalities will go beyond the acquisition and expert application of the biomaterial at the time of surgery or intervention; additional issues include the follow up, the medium- and long-term accessibility of trained staff, support workers, centers of excellence and financial infrastructures (insurance and licensing) to offer ongoing support for people using the biomaterials. Open sourcing of biomaterials to enable more countries to benefit from the technology would be desirable.

3.2 Biopolymers/Plastics

3.2.1 Introduction

Biopolymers are polymers produced by living organisms. Cellulose and starch, proteins and peptides, and DNA and RNA are all examples of biopolymers in which the monomeric units are sugars, amino acids and nucleotides, respectively. Bioplastics or organic plastics are forms of plastics derived from renewable biomass sources, such as vegetable oil, starch or sugar, rather than fossil-fuel plastics, which are derived from petroleum. Some biopolymers are biodegradable: they are broken down into CO_2 and water by microorganisms. In addition, some of these biodegradable biopolymers are compostable: they can be put into an industrial composting process and 90% of the matter will break down within six months. Such biopolymers can be marked with a "compostable" symbol, under European Standard EN 13432 (2000). Many bioplastics lack the performance and ease of processing of traditional materials but are used in order to engage with environmentally sustainable manufacturing practices (Plastics2020Challenge, 2010). Bioplastics rely less on fossil fuel as a carbon source than petroleum-based plastics, introduce fewer net new greenhouse emissions and do not produce hazardous waste residues upon biodegradation. Some bioplastics are used to create disposable items that are compostable and will degrade, while another important use of bioplastics is to create sustainable material alternatives to oil-based plastics. Bioplastics can be made from agricultural byproducts fermenting by natural or engineered microorganisms. The primary applications of bioplastics are food packaging, food waste collection bags and agricultural mulch films to suppress weed growth.

Types of bioplastics currently include:

Thermaplastic starch, the most widely used bioplastic for pharmaceuticals capsules. These tend to be highly biodegradable and among the cheapest. The chemical firm Novamont produces Mater-Bi in Italy (Novamont, 2010) using corn and potato starch. It has capacity for 60 000 t/year. Thermally destructured starch is blended with a fossil-based but biodegradable copolymer to add strength. Blended formulations are increasingly being used to achieve enhanced material properties such as moisture resistance or a gas barrier. The Australian firm Plantic avoids the need for a copolymer by using cornstarch high in amylose, a major natural component of starch. Plantic has a 7500 t/year capacity. The Dutch firm Rodenburg produces its Solanyl bioplastic from waste potato peelings at a 47 000 t/year facility (Solanyl, 2008).

Cellulose-based materials, to produce a thin, home-compostable plastic film. Innovia Films lead this category, with a 30 000 t/year plant for its Natureflex range in Cumbria (Innovia Films, 2010). Cellulose is extracted from sustainably sourced wood pulp and chemically modified before extrusion.

Polylactic acid (PLA), produced from cane sugar or glucose, used in the packaging industry. This is made by Natureworks, a joint venture between Cargill and Teijin, in the Nebraska corn belt (Natureworks LLC, 2010). It will reach its full 140 000 t/year capacity when a second production line opens in April (2009). Cornstarch is fermented to make lactic acid before polymerization to the final PLA product. PLA could, in the future, be made from non-food crops via second-generation cellulosic technology. The lactic acid maker Purac is collaborating with Sulzer Chemtech and the expanded polystyrene producer Synbra to open a small PLA-based "Biofoam" plant in the Netherlands by the end of 2010, with plans to expand PLA capacity to 50 000 t/year (Synbra, 2009).

Fossil fuel-based bioplastics, BASF has the capacity to produce 14 000 t/year of its Ecoflex biodegradable polyester (Treehugger, 2007).

Polyamide 11 (PA11), a biopolymer produced from natural oil used for thermally resistant high-performance materials such as automotive fuel lines but is not biodegradable.

Bio-derived polyethylene, used in packaging and is produced from fermentation of agricultural feedstocks such as sugarcane or corn. It does not degrade but can be recycled and is associated with reduced greenhouse gas emissions

Polyhydroxyalkanoate (PHA) is one of the most promising biopolymers produced by many bacteria (Aldor and Keasling, 2003; Lee and Lee, 2003). PHAs have useful properties and true biodegradability. These properties are similar to those of polyethylene and polypropylene and can be blended to engineer polymers with the desired properties. PHA has experienced a protracted development time to market. Developed by ICI in response to the 1970s oil crises, PHAs have not yet found commercial success. Telles, a joint venture between the biotech firm Metabolix and the agrifood business Archer Daniels Midland, plans to bring its Mirel PHA product to market in late 2010 at about twice the price of conventional plastic (Mirel, 2010). It will be produced by using genetically modified (GM) microbes to ferment cornstarch. A plant with a 50 000 t/year capacity is being built in Iowa. Telles intends eventually to produce PHA from generically modified switchgrass, using the left-over biomass for cellulosic ethanol production. PHAs could be produced by pure cultivation via fermenting using purified sugar as substrates, but this is not economically competitive compared to those produced by chemically synthesis. A mixed bacterial culture production has been developed to produce large amounts of PHAs at lower cost and can utilize a wide range of cheap substrates including industrial and agricultural wastes (Rhu *et al.*, 2003).

Poly-3-hydroxybutyrate (PHB), a type of PHA, is a polyester used to make heat-tolerant, clear packaging film, produced by starch- or glucose-processing bacteria on an industrial scale.

Bio-based non-degradable bioplastics are predicted to take off in the next few years and could overshadow biodegradable materials. The Brazilian firm Braskem is set to start operating a 200 000 t/year plant to make polyethylene from sugarcane ethanol in 2011. It hopes to command a 15–30% price premium over fossil-based polyethylene (Ethanol Business, 2007). Dow Chemical plans to build a 350 000 t/year bio-polyethylene plant by 2012 (ENDS Report, 2009). Meanwhile, Brussels-based Solvay has plans to produce 360 000 t/year of polyvinyl chloride from sugarcane ethanol at its site in Brazil (Solvay Plastics, 2010). Solvay is already the largest PVC supplier to Europe.

SB-based manufacturing processes have the potential to develop a new generation of bioplastics industry that is more efficient and flexible than the current bioplastics industry. It could compete with petro plastics on account of its environmental credentials and materials performance. SB-based techniques may also help develop novel microorganisms that can digest currently indigestible plastics. The technical challenges are formulation development for biopolymer applications to allow biopolymers to be tailored for specific applications, which is the major strength point for plastics expansion. Two key challenges are to produce bioplastics that do not rely upon petroleum as an energy and materials source and to create completely biodegradable bioplastics. Another challenge will be to develop bioplastics that are degradable by anaerobic digestion so that foodstuff does not have to be separated from its packaging for waste recycling processes. Further technical challenges involve using SB methods to produce bioplastics that have consistent properties and are not batch variable, as well as being able to improve the durability, appearance and longevity of bioplastics and biopolymers for the structural materials market. Biopolymer formations with a high moisture resistance are also needed for the structural materials market. Recycling techniques need to be created in order to extend the useful life of biopolymers rather than consign them to biodegradation/composting processes. Also, new formulations are needed to allow tailoring biopolymers for specific applications in line with the market trend in petrochemical polymers. This trend requires the concurrent development of additives, fillers and reinforcing agents that are produced from natural or biodegradable sources that can be incorporated into biopolymers. Biomonomers for the production of traditional polymers for industrial and structural applications are necessary. In this case, the additives used cannot be biodegradable. Finishing processes need to be designed that are compatible with a range of biodegradable materials for a variety of structural uses. Effective SB-based production processes are needed that can operate on a scale large enough to enable the biopolymers market to become competitive with the petro plastics market. In 2008, Natureworks introduced a more efficient fermentation technology, which has reduced carbon emissions by 60% and energy use by 30% during the manufacture of polylactic acid (PLA). The improvements mean PLA can now compete with fully recycled polyethylene terephthalate (rPET) in terms of carbon emissions, with a 14% comparative reduction in CO_2 of life cycle emission. Natureworks claims that PLA's lower density means less material is needed to achieve equivalent strength, so the climate advantages will be even greater on a functional unit basis. The next few years could see the introduction of plant-derived but non-biodegradable bioplastics that are functionally and chemically identical to their oil-derived counterparts. Large-scale plans have been announced for Brazilian bio-polyethylene from sugarcane ethanol. This could directly replace oil-derived material, fitting naturally into existing processing and recycling infrastructure and opening up massive markets that would be inaccessible to biodegradable materials (ENDS Report, 2009).

3.2.2
Economic Potential

The performance of the biopolymers market has been very unpredictable, making it a risky market for investors and difficult for bioplastics producers to penetrate the packaging industry. For example, in late autumn 2007, the supermarket chain Asda announced a "major switch" to bioplastics for packaging (ENDS Report, 2002), but now it does not use them at all.

The market for bioplastics has been growing fast but still remains tiny compared with conventional alternatives; it accounts for 60% of the biodegradable materials market in Europe. The most common end use market is for packaging materials. Japan has also been a pioneer in bioplastics, incorporating them into electronics and automobiles. While use of bioplastics has since grown, today they make up only 0.2% of the total 48 million t/year EU market for plastics, according to the trade body European Bioplastics. The plastics market is valued in tonnes because the unit prices vary greatly according to the market and also depend on what kind of plastic is being traded. Assuming 1 t of plastic has a value of about US$ 1000, the value of the bioplastics market in 2010 can be estimated at US$ 250 billion (London Metal Exchange, 2010). In a 2003 report, continued growth of the plastics market in general has been predicted over the next 10 years (Michael, 2003).

Supermarkets' caution has certainly had an impact on the sales growth of bioplastics, at least in the United Kingdom. ENDS reported that, in 2000, a minuscule 25 000 t of bioplastics were consumed worldwide, but manufacturers boldly predicted a near 10-fold increase in capacity by 2002 (ENDS Report, 2004). That optimistic target was finally reached in 2007, when European Bioplastics estimated a 262 000 t global capacity. It reports that 75 000–100 000 t of bioplastics were consumed that year in Europe. European Bioplastics now expects world production capacity to reach 1 500 000 t in 2011, including 575 000 t of bio-based materials such as bio-polyethylene. The accuracy of these predictions is unclear, and with bioplastics still costing more than their competitors, the price of oil is a key variable. European Bioplastics suggests market potential in Europe alone could reach five million tonnes in 2020, a 10% share of the total plastics market assuming no overall growth. Such dizzying heights would require a growth rate of about 35% per year. A 2005 report prepared by Utrecht University and Fraunhofer Institute for the European Commission's IPTS 3 noted that the technical substitution potential of biopolymers is estimated at 33% of total polymer production. According to the European Bioplastics Association the technical potential could be estimated at 5–10% of the plastics consumption and in the long run much higher. Global interest in sustainable manufacturing processes and products is high and can attract investor sentiment, especially when the conventional plastics industry carries a negative environmental public image. Bioplastics and biopolymers promise a variety of environmental benefits, ranging from renewable resource usage to carbon savings. One example is in applications where plastic packaging is difficult to collect and recycle; here, using biodegradable materials offers an advantage to end users. The limited volume of landfill for the disposal of household waste ensures the continued need for biodegradable packaging materials.

Despite promising environmental credentials, bioplastics were recently unable to compete price-wise with the established petro plastics producers and lacked compelling performance advantages to make up for the price difference. Plastics recycling schemes and incinerators are countering the issue of waste disposal in limited landfill space. The scale of production possible with SB-produced biopolymers is not competitive with the established petro plastics manufacturing processes. The SB-produced biopolymers industry needs access to low-cost materials of consistent quality. The "sustainable" identity of bioplastics is misleading because many products continue to rely on traditional petroleum-based processes. This has the potential to confuse customers and deter sales. It is anticipated that market traction by biopolymers will depend on an expansion in consumer goods and in engineering applications. Thus, growth of the SB-produced biopolymers market will rely on innovation, for example, formulations to improve the performance of structural bioplastics. There are significant market opportunities for such biopolymers if the price of oil remains high and market segmentation is possible for the development of specialty plastics. New market opportunities for the SB-based development of bioadditives and biofillers and an innovative approach to the SB-produced biopolymers markets are very likely to increase the market size. The changing political and economic landscape for waste management should provide a boost for bioplastics because reduction targets are driving investment in incinerators, composting plants and anaerobic digesters. All these approaches are broadly beneficial end of life options for bioplastics because conventional plastics can only be recycled, incinerated or landfilled. The major threat to the bioplastics market is that it cannot become competitive with petro plastics, and government regulations or legislation of new products may make it even more difficult to compete with an already difficult to penetrate established market.

The total built-in capacity for biopolymers in 2009 was around 226 000 t. These include: (i) polylactide acid (PLA; NatureWorks, Galactic, Hycail BV), (ii) polyhydroxyalkanoates such as PHAs, PHB, and PHBH (Biomer, Procter&Gamble), (iii) polymers based on bio-based PDO (DuPont), (iv) cellulose polymers (Innovia Films), (v) epoxy polymers from bio-glycerol and (vi) starch polymers and blends. NatureWorks (Cargill Dow) is the major commercial player, with a PLA capacity of 127 000 t; and Novamont is the major producer of starch polymers and blends, with a capacity of 54 000 t. The total capacity of biopolymers is expected to reach 590 000 t, including a contribution from Braskem's 180 000 t/year of bio-ethylene production and Braskem's/Novozymes 180 000 t/year of bio-polypropylene production in Brazil. Based on this estimate, the share of biopolymers in the total global production of synthetic polymers will still be merely 0.26% (Rao, 2010).

3.2.3
Environmental Impact

Many currently available biopolymers are biodegradable (cellulose and paper are typical examples); many products can be developed using other types of polymers, in particular those containing sulfur. In those cases, there could be a negative impact of sulfur compounds due to various types of toxicity. However, mineralization of sulfur in soil will ease the problem somewhat.

An environmental indicator that is relevant for biopolymers produced on a global scale is greenhouse gas emissions. Agricultural byproducts could supply at least 40% of the current demand for polymers if matured processes were developed. The ability to utilize bio-based feedstock could in itself effect a marked reduction of GHGs. In addition, production processes that have fewer steps and are cleaner could reduce the energy consumption and waste generated from industrial processes.

The convert marginal lands into farms for the production of corn or sugarcane is an excessive environmental cost. Such change of land use also involves the use of large numbers of agricultural machines (tractors, harvesters, tillage equipment, grinders, choppers, etc.) that require energy. In addition, excess chemical fertilisers are added to the environment to increase the harvest yield. Industrialized, large-scale farming to harvest biomass will create more greenhouse gases. Moreover, the use of chemical fertilisers will additionally impact the ecosystem and boost greenhouse gas emissions (Jones, 2009).

Industrial production of biopolymers based on advanced techniques could have an eco-friendly impact on the environment: it increases process efficiency and enables the use of renewable feedstock. This would reduce greenhouse gas emissions such as CO_2, pollute water and air less and preserve the fossil resource.

A series of principles has been proposed (Anastas and Warner, 1998) to assess the "green chemistry" and can be adapted to evaluate the environmental effects of biopolymers:

"1) It is better to prevent waste than to treat or clean up waste after it is formed.

2) Synthetic methods should be designed to maximize the incorporation of all materials used in the process into the final product.

3) Wherever practicable, synthetic methodologies should be designed to use and generate substances that possess little or no toxicity to human health and the environment.

4) Chemical products should be designed to preserve efficacy of function while reducing toxicity.

5) The use of auxiliary substancces (e.g., solvents, separation agents, etc.) should be made unnecessary whenever possible and innocuous when used.

6) Energy requirements should be recognized for their environmental and economic impacts and should be minimized. Synthetic methods should be conducted at ambient temperature and pressure.

7) A raw material feedstock should be renewable rather than depleting whenever technically and economically practical.

8) Unnecessary derivatization (blocking group, protection/deprotection, temporary modification of physical/chemical processes) should be avoided whenever possible.

9) Catalytic reagents (as selective as possible) are superior to stoichiometric reagents.

10) Chemical products should be designed so that at the end of their function they do not persist in the environment and break down into innocuous degradation products.

11) Analytical methodologies ne ed to be further developed to allow for real-time in-process monitoring and control prior to the formation of hazardous substances.

12) Substances and the form of a substance used in a chemical process should chosen so as to minimize the potential for chemical accidents, including releases, explosions, and fires."

3.2.4 Foreseeable Social and Ethical Aspects

The reasons to replace the current methods of biopolymer and plastics production by new ones based on the use of synthetic microorganisms are: (i) to abandon polluting chemical industries, (ii) to create more sophisticated materials having new properties and (iii) to create more easily degradable materials. If abandoning the current, polluting chemical industries is positively received, then public acceptance will mainly depend on the nature of the products. For example, materials produced by microorganisms are not necessarily more easily biodegradable. As in the case of other biomaterials, the risk of dissemination in nature is not high and biosecurity is a non-issue. The sophisticated technologies necessary to produce these materials cast doubt on whether they will be more easily shared than the traditional ones.

Biopolymers are related to a broader issue, namely that of society being faced with an environmental "crisis" and attempting to find new, acceptable ways of dealing with such challenges while living everyday life. Consumer choices will be regarded as being an important behavioral input. Being a good citizen will involve an "environmentalist" code of conduct including recycling or home composting of biodegradable bioplastics. Biopolymers and plastics will contribute to the confusion that already exists around compostable plastic systems because concepts of biodegradability are based on complex government and scientific definitions. The terminology is currently very complicated and the way the messages are delivered will have an impact on consumer choice and recycling practices. For example, we need for clarity about the complete life cycle of biopolymers (e.g., land filling with biodegradable biopolymers can actually result in methane emissions). We also need an increased level of communication about how fast biopolymers really degrade. Consumers should be encouraged to develop an understanding about environmental practices such as home composting, which is an attractive option for bioplastic packaging. For this to become reality, consumers need to know which materials are compatible with garden composting: there is no standard, simple way of identifying them. The above concerns are probably temporary because future waste management systems will no doubt better realize the environmental potential of bioplastics.

A social understanding of the environmental effect of biopolymers and plastics is complicated by a controversial, complex situation, namely the public understanding of and engagement with sustainability. For example, the life cycle picture of bioplastics versus petroplastics is already very complex. This makes it even harder to understand the benefits of adopting new practices or making new purchases. While many bioplastics do appear to offer substantial environmental gains over conventional materials, much depends on whether they end up being land filled, incinerated or composted and how different environmental objectives are weighted. A comprehensive life cycle study of plastic carrier bags conducted for the York-based National Non-Foods Crop Center (NNFCC) indicated that such bags offer environmental benefits but are only effective when mass marketing is coupled effectively with disposal routes based on incineration or municipal composting (ENDS Report, 2009).

Supermarket purchases are central outlets for the widespread adoption of bioplastic packaging. While supermarkets were bursting with enthusiasm a few years ago (ENDS Report, 2010), they are now taking a more considered approach. Sainsbury continues to introduce more bioplastics where it feels it is appropriate, but is now focusing on home-compostable materials. There have been marked improvements in the available formats for compostable plastics, now including films, netting and tags. Nine-tenths of the packaging for the company's organic range is either recyclable or compostable. The hope is that this will drive consumers to their outlets. Tesco has developed a successful standard and logo for home-compostable bioplastic. The chain currently stocks about 70 organic product lines in home-compostable plastic. Marks and Spencer has a more conservative view of bioplastics and has set itself a target to use only recyclable or compostable packaging by 2012. Novamont accepts that bioplastics suit some applications but are not a panacea. For instance, out of date food waste from supermarkets could be sent for composting without spending time and money on removing packaging. Asda's change of policy on bioplastics has been the most dramatic: pricing played a key role in the decision to withdraw bioplastic packaging, which was four or five times more expensive than the conventional equivalent.

The costs versus benefits of regulating waste disposal needs and how this is funded are important considerations. For example, bioplastic manufacturers have little control over the fate of their bioploymers which will be used in different end products. United Kingdom homes, for example, throw away about six million tonnes of packaging made from different types of polymers each year. A government strategy for packaging waste (ENDS Report, 2008) made no mention of bioplastics. Although recycling is technically feasible for some bioplastics, this does not happen because volumes are too low to make recycling economically worthwhile. At the same time, such mixure of plastics are a big headache for conventional plastic recyclers.

Ignorance about environmental practices needs to be addressed with clear messages, guidelines, and code of practice and enforcement options. A 2007 consumer study carried out for the government-backed Waste and Resources Action Programme (WRAP) found consumers were struggling to know where to put bioplastics: many persons thought they could simply recycle them with other plastics (ENDS Report, 2009). Conventional plastics recyclers are very concerned about what bioplastic will mean for them because separating out collected plastic waste greatly increases recycling costs. Contamination problems depend on the bioplastic in question. Contamination is not yet a problem because such plastics are present at such low levels. Note also that optical scanning equipment capable of identifying different polymers is being developed to monitor the composition of waste plastics.

The food/fuel trade-off, in which agricultural resources (food and water) become feedstock for SB-based industrial processes instead of being prioritized for human consumption, is a major global issue. Only a few percent of crude oil is used for plastics production. The problem involving the raw material availability of biopolymers is correlated with the problem of biofuels production. For example, the

United States would need 43% and the European Union 38% of the overall agricultural area to replace 10% of the fossil fuel used for transportation (European Bioplastics Conference, 2007).

3.3 Bulk Chemical Production

3.3.1 Introduction

Bulk chemicals are made in very large quantities by standardized reactions. Since the quantities of chemicals needed are so large, it is important for manufacturers to develop manufacturing processes that are as efficient as possible. Bulk chemical manufacturers have relied on catalysts to achieve this. Typical catalytic materials include proton acids, platinum and other metal catalysts, oxygen, strong acids, acrylonitrile from propane and ammonia. The chemical manufacturing industry gained widespread notoriety in the 1970s owing to unregulated chemical manufacturing practices that caused significant environmental pollution. Since then, the industry has promoted a "green" agenda and invested in environmental practices (Green Chemicals, 2009).

The worldwide trend towards sustainable (clean, green) chemical technologies chiefly consists in:

- ***Minimization of raw material consumption*** (the rise in selectivity due to the choice of the most suitable process, feedstock, optimal operating parameters, highly selective catalysts, recirculation of byproducts, etc.);
- ***Minimization of energy consumption*** (the rise in selectivity, the replacement of endothermic processes by globally exothermic ones composed of endo- and exothermic component processes or processes and operations, a gradual replacement of high energy-consuming distillation or cryogenic separation by adsorption, membranes, etc., advances in the construction of reactors, heat exchangers, process control and steering, etc.);
- ***Raising environmental safety*** (minimization of the amounts of emissions, sewages and wastes, low-waste or zero-waste technologies, increased safety of processes, transport, products and their applications, the utilization, processing, recycling, destruction, combustion or safe deposition of wastes from the main production, sewage processing, electric and heat generators, water processing and post-consumer processing).

To achieve the basic effects of technological sustainability, the following main tools can be used both to advance known technologies as well as to develop novel, more sustainable (clean, green) chemical technologies: primary resources, feedstocks, processes and operating conditions, energy sources and energy management, alternative catalysts, alternative reactor/equipment systems, byproducts, emissions, waste treatment and utilization, choice of the best and safest products

(Tanewiski, 2006). The development of metabolic engineering over the last two decades has yielded novel microbial strains for use in sustainable and cost-competitive bioprocesses that are of interest to the bulk chemical manufacturers. This particularly refers to alternative chemical operations. The impact of these changes shows substantial but not rapid progress. Notable examples include, the production of bulk chemicals such as propanediol and ethanol and biopolymers such as poly(hydroxybutyrate) (PHB) and other poly(hydroxyalkanoates) (PHAs).

The goal for the bulk chemicals industry is to efficiently generate huge amounts of compounds using simple, inexpensive starting materials (e.g., glucose) with minimal environmental impact (cleaner, greener, safer). SB allows the engineering paradigm to be applied to biology in a rational and more systematic way than previously possible. It also expands the scope of what can be achieved this way. The greatest impact will be on the production of more efficient and new enzyme systems that can catalyze bulk chemical reactions. By carefully applying metabolic engineering and SB, this biotransformation potential in the production of "designer" biocatylists can be harnessed to produce chemicals that address unmet clinical and industrial needs. In order to utilize biology to perform chemistry, it is important to develop novel bioprocessing approaches to increase the capacity to control over both the function of synthetic biological systems and the engineering of those systems. Recent efforts have improved general techniques and yielded successes in the use of SB to produce drugs, bulk chemicals and fuels in microbial platform hosts (Carothers, Goler, and Keasling, 2009). While traditional biotechnology has had some notable achievements in several of these areas, they have generally been slow and expensive to develop. A typical approach in bioengineering is to develop cells or molecular modules with new functions based on the empirical and/or evolutionary processes that may involve screening candidate libraries. These processes are usually composed of complex biological moduels and difficult to optimize. Today's biotechnologist therefore needs to master a very broad array of complex technologies in order to achieve a goal. By potentially re-organizing biotechnological development in line with the principles of SB, research and development are likely to proceed much faster and in a much more rational way. New design rules will probably aggressively tackle the problems encountered by the traditional empirical approach. Because of its rational, knowledge-based approach to biological design, SB could allow such goals to be attained more quickly and inexpensively. It may also enable developments that current procedures cannot deliver. This include the coordination of complex sequences of enzymatic processes in the cell-based synthesis of useful organic compounds. SB therefore represents an opportunity to develop a second generation of enzymes to specifically manufacture large quantities of highly purified chemicals and processes that are environmentally friendly (NEST Report, 2005). As the world's fossil fuel reserves are waning, the chemical industry needs a new raw materials base. SB might be the tool that enables this change. Microorganisms that efficiently produce the bulk chemicals that supply today's raw materials will be redesigned. They will then be able to make increasingly complex chemicals from these ingredients using even more intricate combinations of synthetic pathways.

Change takes place more slowly in bulk versus fine chemicals production. This will require a widespread, integrated, well-coordinated set of new approaches to develop SB-based manufacturing techniques on the scale necessary to manufacture chemicals in bulk. A portfolio of technological changes must be considered: novel or improved processes, alternative routes, alternative feedstocks (including renewables), advanced and novel catalysts (including biocatalysts), nanotechnology and nanomaterials, membrane and catalytic membrane technologies, microchemical engineering, advanced media (including green solvents, ionic liquids, supercritical fluids), novel reactors and advanced process control equipment (monitoring, automatization, analytical techniques), advanced separation techniques, new sophisticated materials and environmentally benign green products (including functional and biocompatible, intelligent, electronic and energetic, polymers).

The first SB-based bulk chemical processes are just starting to appear on the market. They are mainly chemical substrates for the production of biofuels and biopolymers.

Metabolic Explorer and IFP Inc. are currently accelerating the scaleup of Metex's proprietary technology for producing 1,3-propanediol (PDO). The biological process uses glycerin, a byproduct of biodiesel production, as a feedstock to make PDO. The latter is a bulk chemical used to make polyester fibers, coatings and plastic films. The Metex pilot production facility can be used to make any of the company's proprietary products, including 1,2-propanediol (MPG), PDO and butanol (Biopact, 2007).

Wacker's 500 t/year ACEO bio-acetic acid pilot plant in Burghausen has already been operating for the past six months using three bio-based acetic acid routes (ICIS Green Chemicals, 2010):

ACEO process. Involves a biomass feedstock being converted to ethanol (using yeast) and producing acetic acid via a gas phase oxidation process.

Fermentation to butanediol. Ferments biomass feedstock using bacteria to butane 2,3 diol, which can be then dehydrated to produce methyl ethyl ketone (MEK) or directly produce acetic acid via gas phase oxidation. Acetic acid from MEK is also possible via gas phase oxidation.

Homoacetate fermentation. Ferments biomass feedstock to acetate/acetic acid using bacteria.

The ACEO pilot plant can be expanded depending on the economic situation because bio-based chemicals (in general) are mostly competitive if crude oil exceeds US$ 80/bbl. The ACEO process produced more than 90% bio-acetic acid yield and is being licensed to customers on an individual basis. The next logical step is to produce bio-ethylene for Wacker's polymer products, which include vinyl acetate (VAE), vinyl acetate monomer (VAM), polyvinyl acetate (PVA), polyvinyl alcohol solutions, dispersions and dispersible polymer powders, vinyl chloride copolymers and terpolymers (among others).

3.3.2
Economic Potential

The current market value is US$ 77.33 billion (2005), with a combined annual growth rate of 9.2% (Piribo, 2010). There is a significant market opportunity that is catalyzed by yo-yoing oil prices, new regulations and pressure from consumers and retailers for "green" chemistry (New Scientist, 2010).

The Dow Chemical Company proposes to address more than US$ 20 billion of market opportunities in the development of innovative product solutions for salient business opportunities in the areas of energy, consumerism, transportation and infrastructure, and health and nutrition (Dow, 2010).

Estimates of the percentage market share of SB-based applications are high – from 10% in 2020 to 50% in 2050 (Dornburg, Hermann, and Patel, 2008). White biotechnology (the use of fermentation and enzymatic processes in industrial processes) is receiving increased attention in industry and in government policy and therefore gaining momentum in the European Union. White biotechnology will probably play a key role in the cleaner production of bulk chemicals because it helps save resources and reduces the environmental impacts of chemical productioncan also help exploit biomass as feedstock for industrial and energy production.

There is overwhelming global political and economic demand for more environmentally friendly and sustainable chemical manufacturing processes and products. However, this established market is difficult to penetrate, and change within the industry is slow: the development of clean technology within this field remains a significant scientific challenge. A huge opportunity exists for SB-based manufacturing practices to develop a significant stake in the clean technology market sector by producing bulk substrates that feed biofuel and biopolymer production. Important growth areas are the development of innovative product solutions to address salient business opportunities in the areas of energy, consumerism, transportation and infrastructure, and health and nutrition. Integrating SB with traditional chemistry will open up further new market opportunities. One example is the production of smart chemicals on biopolymer materials for surface coatings that are both structural and functional. Threats to this market are that the industry itself is conservative and the end users are not willing to trade off clean and sustainable technologies for performance regardless of the business niche, product line or their geography. Additionally, regulatory standards and controls may hamper new sustainable practices, products and manufacturing processes.

At present, over 90% of all organic chemicals are made from fossil resources. Most of the bio-refineries are on a relatively small scale and are currently not economically feasible. Faced with increasing oil prices, the chemical industry is seeking alternative sources for bulk chemicals derived from petroleum and other fossil sources. At a raw oil price over US$ 50 per barrel, the bulk conversion of renewable raw materials to chemicals will become practicable and holds significant commercial potential. Besides the oil price, the technological progress in SB boosts the economic viability of large-scale production. A given bio-based bulk chemical product is considered economically viable if its production cost plus profits (PCPP) is lower than the market price or the PCPP of its petrochemical counterpart (Hermann, Blok, and Patel, 2007; Hermann and Patel, 2007). The economic assessment comprises direct operation costs, additional costs and capital charge. The direct operation costs are composed of variable costs (feedstock, auxiliaries/catalysts, byproducts, utilities, waste treatment) and fixed costs (supplies, labor). Additional costs include taxes, insurance fees and plant overhead as well as an allowance for marketing, administration and R&D. The capital charge represents the total of depreciation and profit, which is calculated by multiplying the total fixed capital with a fixed percentage.

Using the bio-based production of 1,3-propanediol (PDO) as an example, the PCPP from four generic bioprocess routes are analyzed for high (400€/t) and low price (70€/t) of sugar fermentation; these are compared to the PCPP of the petrochemical production route at crude oil prices of $ 25/bbl and $ 50/bbl (Figure 3.2). PDO is a product that is currently manufactured on a relatively small industrial scale and sold at high prices for niche applications. Both the petrochemical and the bio-based routes of PDO are implemented at an industrial scale. Shell currently produces PDO from petrochemical feedstock, while DuPont is building a PDO production plant based on fermentable sugars from maize starch. The four generic routes are based on aerobic fermentation of fermentable sugar to PDO. Routes 1 and 2 represent current technology, with downstream processing by distillation. Routes 3 and 4 represent future technology: the separation of PDO by pervaporation, which has been done on the laboratory scale (Li *et al.*, 2001). The current industrial route of DuPont is by sugar fermentation. The petrochemical route is to produce PDO via hydroformylation of ethylene oxide. The PCPP of the different routes to PDO has been calculated. The PCPP values of route 1 and the current

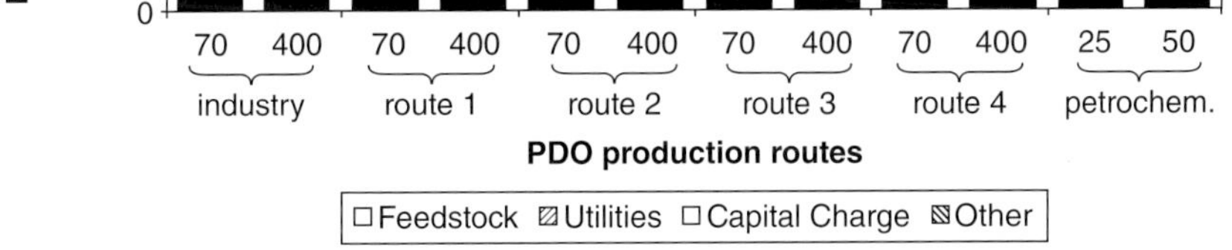

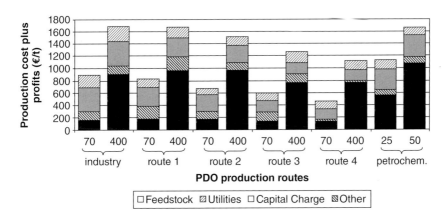

Figure 3.2 Composition of 1,3-propanediol (PDO) costs. Industry: bioprocessing route by DuPont using maize starch. Routes 1–4: generic routes. Petrochem: chemical route of Shell using petrochemical feedstock. Costs: 70 for low fermentable sugar price, 400 for high fermentable sugar price, 25 for crude oil price of US$ 25/bbl, 50 for crude oil price of US$ 50/bbl (Hermann and Patel, 2007).

industrial route are matched well, corroborating the generic approach. Compared with the PCPP of petrochemical PDO, the PDO produced by future technology (route 4) is competitive with petrochemical PDO, even at a sugar price of up to 400€/t or crude oil prices of US$ 25/bbl. Moreover, at high fermentable sugar prices (400€/t), feedstock costs alone may be higher than the total PCPP for low sugar prices (70€/t). At low sugar prices, the dependence of bio-based PDO on feedstock prices is much weaker compared to petrochemical PDO. Generally, at high sugar prices, the contribution of feedstock costs of all routes of bio-PDO are higher than the petrochemical processes at US$ 25/bbl. However, when the oil price increases (US$ 50/bbl), the cost of feedstock of bioprocess becomes competitive with the petrochemical routes (Hermann and Patel, 2007).

Another bulk chemical example is succinic acid. It is currently made mainly from fossil-derived maleic anhydride and functions as the basic four-carbon backbone for a wide range of products including pharmaceuticals, coatings, polymers and resins. Succinic acid is sold at US$ 3–5/kg and the price depends on oil prices. Commercial production of "bio-succinic acid" involves the fermentation of renewable feedstocks such as sugars by microorganisms. Succinic acid is currently only a niche product, with the 30 000 t produced each year creating a market worth US$ 225 million. The market research firm Frost & Sullivan estimates the market will expand sixfold to 180 000 t by 2015 due to the introduction of bio-succinic acid. Some chemical companies believe that this new compound could create a market at least ten times the current one, and this bulk chemical is set to become a commodity chemical.

3-Hydroxypropionic acid (3HPA), produced mainly in the petrochemical industry by oxidation of propylene, can be recovered and transformed into chemical derivatives. One such derivative is acrylic acid – a high-value, high-volume chemical that goes into a broad range of materials including plastics, fiber, coatings, paints and super-absorbent diapers. To date, the major source of acrylic acid is crude oil refinement. Almost half of the 3.1 million tonnes of crude acrylic acid produced annually (as data indicated in 2005) is converted to glacial acrylic acid for superabsorbents. The latter is commonly used in personal care items with consumption of more than 1 million t/year. A process developed by Novozymes and Cargill will be able to convert sugar to 3HPA by fermentation of glucose or of another carbohydrate source. This conversion is a multi-step enzymatic reaction within the engineered microorganisms. The acrylic acid market appears to be growing approximately 3–4% annually (Cargill, 2008).

Solvay has investigated an additional US$ 135 million investment program to expand and increase its vinyls production plant in Santo Andre, Brazil. This second stage of expansion involved building an integrated plant to produce ethylene with ethanol originating from sugar cane. Ethylene will be used as feedstocks to produce polyvinyl chloride (PVC), while supplemented with chlorine. PVC is a thermoplastic material currently derived from petroleum feedstock. There are a wide variety of applications of PVC: automotive, construction, chemical industry, consumer goods, electrical equipment, medical devices, packaging, water transport and environmental applications. The plant would then have an installed capacity of 360 000 t/year of PVC, 360 000 t/year of vinyl chloride monomer (VCM), 235 000 t/year of caustic soda and 60 000 t/year of bio-ethylene.

3.3.3 Environmental Impact

The depletion in fossil resources and the climate change associated with CO_2 emissions are the main driving force behind the development of alternative resources for bulk chemicals, namely the replacement of fossil resources with renewable feedstock that will be carbon neutral as well. A full life cycle assessment (LCA) usually includes calculating a range of environmental impacts such as acidification, eutrophication, particulate emissions, human toxicity and environmental toxicity. In most LCAs dealing with industrial processes, the overall environmental impact is calculated based on non-renewable energy use (NREU), greenhouse gas emissions and land use. NREU represents a straightforward and practical approach because many environmental impacts are related to energy use. NREU values represent the cumulative energy demand for the system frm cradle to grave. Greenhouse gas (GHG) emissions are increasingly important due to their roles in climate change. Such emissions are calculated in CO_2 equivalents and consist of emissions from the system in the form of CO_2 or CH4, as well as nitrous oxide (N_2O) from fertilizer in biomass production. Land use refers to agricultural land use only and will gain importance in the future because of the growth of land requirements for bio-based energy, liquid biofuels, bio-based chemicals and food and feed production. The BREW study funded by the European Commission FP5 GROWTH program has conducted investigations on the medium- and long-term opportunities and risks of the biotechnological production of bulk chemicals from renewable resources. Quantitative analysis on the environmental impacts has been done on 21 chemicals; all are environmentally attractive regarding non-renewable energy usage and GHG emissions. The environmental assessment is performed with BREWtool, a modified Excel program following the principles of a LCA (ISO 1997–2000). The cradle to factory system is composed of five subsystems:

1) Agriculture, biomass pre-treatment and conversion;
2) Extraction and processing of non-renewable energy;
3) Bioprocess;
4) Management of process waste;
5) Management of post-consumer waste.

The study showed that, with current technology and using maize as feedstock, 30% of the cradle to factory non-renewable energy saving has been achieved; this saving is 75% if lignocellulosic feedstocks are used and 85% if future technology is applied (Patel *et al.*, 2006). Hermann, Blok, and Patel (2007) assessed the environmental impacts of producing bio-based bulk chemicals. The quantified GHG savings potentials on several bulk chemicals are calculated as listed in Figure 3.3.

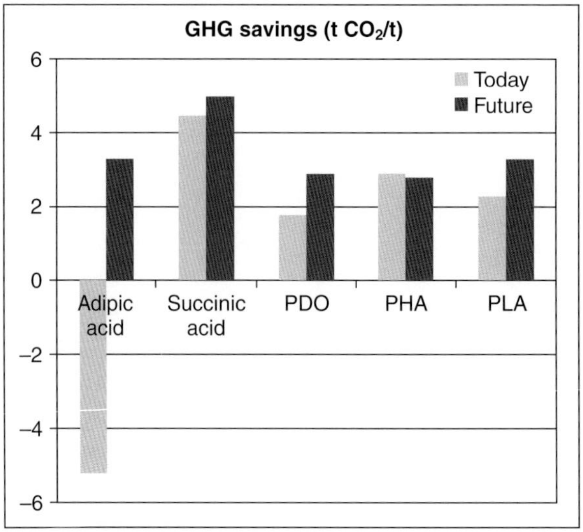

Figure 3.3 Potential GHG savings of five bio-based products compared with conventional petrochemical-based production, using corn starch as feedstock (cradle to grave; Hermann, Blok, and Patel, 2007).

3.3.4 Foreseeable Social and Ethical Aspects

Society would positively receive the disappearance of huge chemical plants, with their inherent risk of manipulating large amounts of dangerous substances at high temperature and high pressure. It would welcome smaller plants, functioning at room temperature and normal pressure (Keasling, 2007; Carothers, Goler, and Keasling, 2009).

Due to the confinement, the risks of the new technology appear limited. Nonetheless, the situation is probably more complex than the previous simplistic presentation suggests. The use of synthetic organisms will not abolish the production of compounds toxic to humans – this depends upon the reactions that will be used and the nature of the final products. The reliability of the process, and the necessity to purify and check the nature of the products, will not always dramatically reduce the size of industrial plants.

Biosecurity will increase because the present huge chemical plants are clear targets for terrorists. Nevertheless, the benefits will have to be estimated case by case.

Over the long term, it is possible to imagine that these new procedures will allow local production in small plants, a benefit for many countries. This level of technical simplification is not a near-future scenario: it will require huge efforts in technological development.

Sustainable development is a humanitarian concept aimed at improving the quality of life and the welfare of humanity equitably, meeting the needs of future generations under the conditions of limited world resources and the far-reaching effects of current industrial activity. Implementing new technologies will require fundamental changes in economic structure production and consumption profiles, technologies, institutions and organization. The role of changes in technology is particularly crucial and will serve as a key driver for the whole transformation of developmental models across society in general.

The ethical problems associated with the chemical industry relate to constant risk and responsibility for chemical processes, chemical products and their safety. Safety regulations would be needed to protect communities from spills of microbial cultures from SB-based manufacturing units. The cost/benefit of manufacturing more environmentally friendly products needs to be weighed against the risks of developing a new and untried technology at such a scale.

It will probably be necessary to create new agencies or regulatory bodies to deal adequately with the potential adverse effects of new technologies. Federal regulatory agencies already suffer from under-funding and bureaucratic ossification. New thinking, new laws and new organizational forms are necessary. Many of these changes regarding an altered manufacturing base will take a decade or more to accomplish, but it is urgent, given the rapid pace of technological change, to start thinking about them now. In the twentyfirst century, governments face emerging technologies that enable common materials and chemicals to be manipulated in sophisticated ways at the nanoscale; these may pose unknown and not yet understood risks to human health and the environment. These novel materials improve performance of the full range of imaginable products, from cosmetics to car batteries to cancer treatments, but little is known about the risks they may pose and there are many questions concerning their life cycle impact.

An important consideration is the risk of bioterrorism versus freedom of public access to biotechnology. Biotech hobbyists and biohackers are already engaged in open sourcing biotechnology (Jeremijenko, 2004). Their practices raise questions about the distinction between those with a keen interest in biological phenomena and others with malicious intent.

Ethical questions relating to the use of SB at the bulk chemical scale for weaponisation and militarization for the purposes of national security need to be posed. This includes the creation of "biological deterrents" and the development of combat biological weaponry.

Widespread global inequalities are expected in terms of access to SB-based manufacturing processed from the perspective of economics, manufacturing infrastructure and geography. The developed world will benefit and the developing world will lose out particularly in the competition for land and water used for the biomass production that underpins a SB manufacturing industry.

3.4 Fine Chemical Production

3.4.1 Introduction

Fine chemicals are pure, single chemical substances that are commercially produced with chemical reactions into highly specialized applications. Fine chemicals can be categorized into:

Active pharmaceutical ingredients (and their intermediates) including vitamins (compounds that are necessary for normal metabolism but not synthesized by the body) and chiral compounds (specific activity conferred by optical chemical structure);

Biocides (pesticides and antimicrobials);

Specialty chemicals for technical applications (flavours, fragrances and pigments).

The production of fine chemicals is expensive, generates significant wastage and requires a higher research investment per kilogram (when compared with bulk chemicals production). This is due to the frequently changing chemistry, the flexibility required to produce a large range of chemicals and the relatively small volumes needed. The fine chemicals industry is an early adopter of new technologies. Biocides and specialized chemicals are developed more slowly than pharmaceuticals. Fine chemicals can be produced in industrial quantities and are sold on a price per mass basis. There is current demand for the fine chemical industry to produce more environmentally friendly processes and products (Carpenter, 2001). Catalysis, for example, plays a key role in the discovery of novel chemical routes for the production of fine chemicals. Accordingly, the importance of biocatalyst technology, as a part of a broader "chemical biotechnology," is increasingly being recognized as an important tool for a chemical synthesis which more efficient and product-specific than chemical catalysis. Biocatalyst R&D aims to produce chemicals by novel approaches to reduce cost of material, water and energy consumption and pollutant dispersal by 30% in the next two decades. The biocatalysts in industrial scale production include the production of high-fructose corn syrup, aspartame and some pharmaceucticals such as vitamins, semi-synthetic penicillins and anti-tumor components. Enzymatic methods are now commonly used to make chiral intermediates, which is all-important in the field of health care products and vitamins. Fermentation and enzyme-based synthetic catalysis are comparable or better than traditional synthetic methods of producing optically active pharmaceutical intermediates. Vitamin B12, for example, is produced for the commercial market by bacterial synthesis alone. SB creates the manufacturing platform for a new generation of environmentally friendly, profitable and increasingly diverse group of fine chemicals. A SB manufacturing approach helps establish a rational and engineering-based synthesis of fine chemicals, industry-based enzymes and complex pharmaceuticals.

A SB approach enables the development of novel bacterial enzymes for the production of a greater range of fine chemicals than is currently possible using chemical catalysts. Examples include greater enzymatic chiral resolution (a prime application for new biocatalysts) as well as the development of new approaches and products such as multi-enzyme pathways for the *in vitro* production of complex, novel, fine chemicals (e.g., unnatural monosaccharides). SB-based approaches enable the development of biocatalysts that are better, faster and cheaper than current chemical catalysts. This approach also helps generate biocatalysts that have increased temperature stability, activity and solvent compatibility for the production of fine chemicals. The approaches also enable the development of novel reaction routes for catalytic transformation of renewable biomaterials. The ultimate goal is to transform affordable material such as sugars, terpenes, natural oils and fats into high value-added products using an environmentally friendly process. SB techniques provide the technical basis for "metabolic engineering" or "multi-enzyme" pathways. This may involve the fine tuning for the exiting drugs to improve their therapeutic properties and to produce low or no side effects for patients. A direct extension of this is the production of new drugs that are based in the known therapeutic properties of certain plants. By the rational design using SB approaches, properties of the candidate drugs can be optimized and side effects reduced. A major development in SB over the next ten years is likely to be the realization of personalized drugs. SB techniques will be used to engineer new types of pesticides. These will be environmentally friendly by very specifically targeting organisms and by having a lifespan in the ground consistent with their function.

Technical challenges include the limited knowledge of enzyme/biocatalyst mechanisms, for exampleof metabolic pathways for secondary metabolites and pathway interactions. The range and numbers of types of enzymes employed in these new manufacturing processes need to be optimized (i.e., expand SB engineering toolboxes to include isomerases, transferases, oxidoreductases, lyases, ligases, synthetic supramolecular systems and chiral catalysts capable of high selectivity for a variety of asymmetric transformations). It is also essential to reduce the time required from research and development of these new techniques to market and implementation. Novel enzymes derived from extremophiles represent an area of high research and commercial interest. These enzymes have gained attention because they evolve under circumstances that can provide activity over a broader range of conditions. Enzymes (native or modified), along with microbial catalysts, are combined to discover new biologically active molecules or to improve lead candidates for pharmaceutical discovery. One aspect is the development of functional platforms based on substrate, milieu, reaction and so on. This will make it possible to rapidly screen and develop biocatalysts for tailored use and reduce the time from research and development to manufacturing and productisation. Another aspect is the elimination of the typical difficulties or operational limits encountered in biocatalysis, such as temperature, pH, product inhibition, slower rates, robustness, durability (including under immobilized conditions) and processing ability in dilute aqueous product streams. This must all be accomplished while maintaining the advantages of high specificity (reaction precision) and multi-step processing (better yields). SB-based manufacturing approaches should address reducing the impacts on water, materials and energy consumption and help reduce toxics and pollutant dispersion. Impacts on carbon management should be positive, that is, help reduce carbon emissions and possibly participate in carbon sequestration. There is a current and rapidly developing market for the production of fine chemicals using SB-based production methods. The anti-malarial artemisinin is currently being produced using SB techniques. In the next five years it will go into full production, with a worldwide impact on malaria. The chemical is being produced in fermenting vats (Hale *et al.*, 2007). The average total cost for the development of a new medicine is US$ 800 million (DiMasi, Hansen, and Grabowski, 2003). The Artemisinin Project, funded by a US$ 42.6 million grant from the Bill and Melinda Gates Foundation, creates a new partnership. It comprises Amyris Biotechnologies, the Institute for OneWorld Health and the University of California, Berkeley. The goal is to inexpensively produce the antimalarial drug artemisinin through a new fermentation process. This San Francisco Bay Area-based public–private partnership is applying a combination of SB, industrial fermentation, chemical synthesis and drug development expertise to a very specific need in the developing world.

3.4.1.1 **Vitamins and Pharmaceuticals**

Ascorbic acid is produced by microbial fermentation (Bremus *et al.*, 2006). Microbial production is the only source of vitamin B12, and riboflavin (vitamin B2) is the only other vitamin that is manufactured microbiologically to any significant extent. None of the fat-soluble vitamins are produced industrially by microbiological methods, but one compound – beta carotene, which is converted by animals into vitamin A – can be prepared by microbial synthesis.

On 6 April 2010, the Autorité des marchés financiers (AMF, the French stock market regulator) had approved the company's prospectus (reference number 10-083) relating to its planned initial public offering on the Alternext market of NYSE Euronext Paris. The company Deinove is seeking to raise around € 12 million in fresh equity (excluding the potential exercise of the extension and over-allotment options). Deinove designs and develops breakthrough technologies by leveraging the exceptional natural biodiversity and robustness of the *Deinococcus* bacterium (which appeared on Earth over three billion years ago) to elaborate innovative industrial processes. This includes the production of bioethanol from all types of biomass and biomass waste (Deinove, 2009). "The IPO will accelerate the industrial deployment of our Deinol bioethanol program with Tereos and the development of our Deinochem (green chemistry) and Deinobiotics (antibiotics) projects. Moreover, stock market listing will raise our international profile with potential industrial partners," emphasized Deinove CEO Jacques Biton.

The company Amgen pioneered the development of novel and innovative products based on advances in recombinant DNA and molecular biology. More than a decade ago, Amgen introduced two of the first biologically derived human therapeutics, EPOGEN® (Epoetin alfa) and NEUPOGEN® (Filgrastim). These became the biotechnology industry's first blockbuster medicines (Amgen, 2005). These biosimilars (biopharmaceuticals that are biologically similar to an existing medicine) are manufactured using genetically engineered bacteria to produce genetic recombinant sequences for erythropoietin alfa protein. This protein stimulates red blood cells in patients with kidney failure. This genetic sequence is then used to transform Chinese ovarian hamster cells to produce the protein in pharmaceutical quantities.

3.4.2 Economic Potential

In 2006 the fine chemicals market was estimated at US$ 80 billion, with a combined annual growth rate of 10–20% (Table 3.2). Although the global pharmaceutical industry is still expanding, sales growth was expected to slow to 5–6% in 2008 (C&EN, 2008). There is current optimism for the growth of the fine chemicals market, which has grown from about US$ 56 billion in 1999 to about US$ 80 billion in 2006 (C&EN, 2008). About two-thirds of the growth was from pharmaceuticals, while the rest came from agrochemicals and other uses. Optimism prevails in the industry, with anticipated growth rates. In a survey at the Informex custom chemicals trade show, San Francisco, 63% of the people interviewed, involved with fine and custom chemicals, thought industry sales would grow 5–10%. In 2005, the research analysts from Frost & Sullivan predicted revenues in the North American fine chemicals sector would grow from US$ 21.29 billion (€ 17.78 billion) in 2005 to reach US$ 28.62 billion in 2011, driven by the production restructuring announced by pharmaceutical majors (Roumeliotis, 2006). Market performance has already exceeded these predictions.

Table 3.2 Value and projected growth rate of different areas of the fine chemicals market.

Fine chemical sector	Market value (US$)	Projected growth rate (%)
Pharmaceuticals (blood disorders major market)	38.4 bn	4.1–7.7
Green chemistry (soaps and detergents) in fine chiral market	1.25 bn (2006)	7.2–14.2
Fine chiral	16 bn (2006)	13.2
Food (e.g., vitamins)	10 bn (2003)	7.2
Total	80 bn (2006)	5–20

The major growth areas in the fine chemical market are the health-related fields, which have the biggest commercial impact. The price of a kilogram of artemisinin, the key ingredient in the current preferred malaria treatment, artemisinin combination therapy (ACT), ranged from US$ 1200 in 2006 to US$ 150 in 2008. The price is expected to increase due to concerns over a shortage because of poor weather and limited cultivation. A subsidy introduced at the top of the private sector supply chain can significantly increase usage of ACTs and reduce their retail price to the level of common monotherapies. Additional interventions may be needed to ensure access to ACTs in remote areas and for poorer individuals. Such people apparently seek treatment at drug shops less frequently (Sabot *et al.*, 2009). If the project is successful, it could produce enough artemisinin to treat 200 million people out of the approximately 500 million who contract malaria annually, company officials said. Sanofi-Aventis stated that the partners hope to begin selling the product in 2010 (Global Business Coalition, 2008).

Other growing markets include commodities (soaps and detergents) that embrace "green chemistry". The driving force is environmental and energy concerns (U.S. Department of Energy, 1999). Worldwide, the market for chiral fine chemicals sold as single enantiomers was US$ 6.63 billion in 2000 and was expected to grow at 13.2% annually to US$ 16.0 billion in 2007, according to a study by the market research firm Frost & Sullivan, London (Roumeliotis, 2006). The pharmaceutical industry is the engine that is driving this strong growth, accounting for 81.2% of the total. The remaining US$ 1.25 billion is divided among such uses as agricultural chemicals, electronics chemicals, flavors and fragrances. The United States is the biggest consumer of enantiomeric fine chemicals, contributing to a total North American share of US$ 3.98 billion, or 60% of the total. European and Asian consumption of enantiomeric fine chemicals is not expected to grow as fast, with the North American share rising to 66.9% of the market in 2007, or US$ 10.7 billion (Shafaati, 2007).

The sale of single-enantiomer compounds made into the pharmaceutical formulations that people actually consume is even more impressive. The worldwide market for dosage forms of single-enantiomer drugs was US$ 123 billion in 2000, up 7.2% from US$ 115 billion in 1999, according to data developed by Technology Catalysts International, a chiral and fine chemicals consulting and research company. That data suggests tha respiratory, gastrointestinal, ophthalmic and cardiovascular drugs will contribute to strong growth; for example, 72% sales growth of the respiratory drug montelukast by Merck (Whitehouse, N.J.) is expected (Shafaati, 2007).

In 2003 the market for nutritional supplements, a market of vitamins, carotenoids and other fine chemicals to the feed, food, pharmaceutical and personal care, was worth US$ 10 billion through all channels, including mass-market, health/natural and direct (Global Information Inc, 2004).

The economic and industrial potential of SB-based production of fine chemicals is potentially very significant. It promises to reduce the prices of vital medicines, develop new ones and create a more environmentally friendly chemical industry.

Developments in this field will profoundly influence the world economies, medicine and the environment.

The market of SB-derived fine chemicals is at an early stage of growth, and those first in the field will establish themselves as market leaders. The goals: (i) greater amounts of more effective product made more cheaply and (ii) new products to create new market opportunities. This will require the development of foundational technologies that make the engineering of SB-produced fine chemicals more straightforward and reliable. The Artemisinin Project, for example, will provide a benchmark in setting the future market potential for new drugs: the long research and development cycles from scientific breakthrough to commercialization of products. As demonstrated by the Artemisinin Project, effective coordination and communication between academics, companies and national agencies is needed to produce an effective SB manufacturing infrastructure for market growth. This is because the chemical industry is both diverse and complex, and the production of fine chemicals will require appropriate policies and logistical structures to fully develop the manufacturing practice. An entirely new, environmentally friendly chemical industry with relevance to the developing countries relying on SB-based practices is a possibility. Why? Because societal forces are demanding new products, new technologies and new ways of living. Here, government, regulatory and cause-driven factors are drivers of change, particularly with respect to environmental issues and industrial sustainability. European markets will face the most competition, particularly from Asian companies and also from fine chemical producers in nonSB-based growth areas, for example highly potent active pharmaceutical ingredients (HPAPI) and hazardous chemistry markets. Additional threats are the slow development of products and the lack of getting to market. This lowers profitability for investors and reflects the immaturity of the field and lack of appropriate infrastructure. Regulatory risks are also a possible threat to the development of the market because the fine chemicals manufacture of pharmaceuticals and intermediates needs to conform to the strict Good Manufacturing Practice standards, and is monitored by the food and drug authorities (as are some vitamins), particularly the United States Food and Drug Administration (FDA).

3.4.3
Environmental Impact

The establishment of rational and engineering-based synthetic biotechnology processes and protocols for the synthesis of fine chemicals, industrial enzymes and complex pharmaceuticals will have a marked impact on the environment. SB will make a significant contribution to the development of the expanding Green Chemistry which, as opposed to Gray Chemistry and particularly Chlorine Chemistry, focuses on environmental issues. In the Green Chemistry Resource Exchange (GreenChemEx) – a database created by the Green Chemistry Institute (GCI) of the American Chemical Society (ACS) – Draths and Frost[1)]

1) http://www.greenchemex.org/?module=resources.edit&id=38&fs=1

report that many conventional chemicals, among them chlorine, could already be replaced or eliminated by using microbes as environmentally benign synthetic catalysts. More specifically, laccases (a group of oxidative enzymes) are gaining interest because their enzymatic action includes oxygen radicals; these can also be used for bleaching processes, replacing chlorine and ozone (Riva, 2006).

Case studies performed by the Öko Institut show that CO_2 reductions of 17–65% could be achieved from fine chemicals derived from renewable resources (EuropaBio report, 2003). Fine chemicals are largely produced in chemical plants, known to be fairly polluting. For example, they use oil derivatives as starting materials (while engineered biological processes are based on biomass) and a variety of catalysts, in particular toxic metals. This would not be the case with SB approaches, where pollution would mainly come from the chemicals themselves (if the final products are toxic substances by nature) and from biomass production. The major reason why biological process is seldom used is that the yield of biological processes in terms of production of chemicals is usually low. This is particularly worrying for low added value chemicals (even when labeled fine chemicals). SB might considerably increase production compared with standard biological processes (but this still has to be demonstrated). SB would presumably alleviate much of the environmental concerns raised by chemical industry. The positive potential of biosynthesis in environmental issues is enormous. Many synthetic vitamin production processes are being replaced by microbial fermentations, for example the bio-production on vitamin B2. Traditionally, a complex eight-step chemical synthesis was applied to produce vitamin B2. A new bioprocess developed by BASF reduces production to a one-step process by fermentation on renewable feedstock. The final product can be easily recovered as yellow crystals from the fermentation (BASF, 2005). This bioprocess reduces costs by 43% and reduces environmental impacts by 40% (30% reduction in CO_2 emission, reduction in hazardous substances).

An example of an antibiotic bioprocess is the production of cephalexin via a methodology developed by DSM. The traditional chemical process of this antibiotic involves 13 steps. The new route, designed by metabolic engineering, is based on fermentation of an intermediate and enzymatically linked to a side chain. This new route uses less energy (65% reduction) and less input chemicals (65% reduction), and generates less waste (OECD, 2001). Another example of fine chemical production is adipic acid, the organic compound with the formula $(CH_2)_4(COOH)_2$. It is the most important form of dicarboxylic acids, a precursor of nylon produced in amounts of about 2.5 billion kilograms annually. Current commercial processes to produce adipic acid involve two stages: (i) production of cyclohexanone or cylohexanol or a mixture of both (i.e., KA – K for ketone and A for alcohol) and (ii) subsequent conversion to adipic acid. A novel bioprocess route developed by Verdezyne Inc. produces adipic acid from an alkane feedstock by a fermentation process using a yeast microorganism. A metabolic pathway is under investigation to allow the microbes to utilize sugar and plant-based oil feedstocks as well as alkanes. The benefit of a feedstock-flexible process is the ability to maintain a

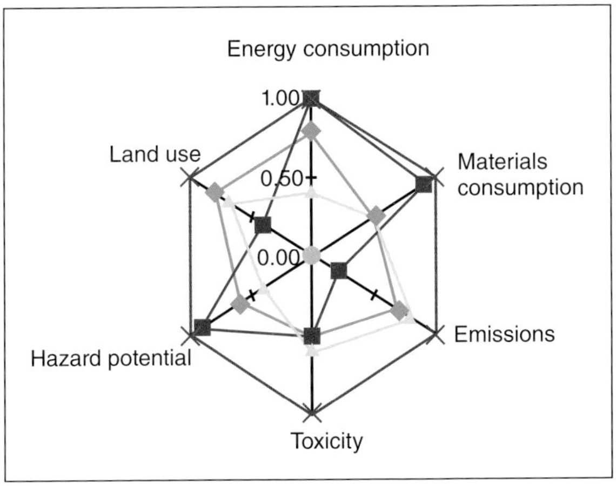

Figure 3.4 Parameters of the eco-efficiency analysis (source: BASF, 2010).

sustainable cost advantage over the possible volatility of feedstock. In addition, the fermentation process will reduce greenhouse gas emissions compared to the traditional petrochemical production of adipic acid (AllBusiness, 2010).

To assess the benefits of "green chemistry", BASF has developed an in-house eco-efficiency analysis to evaluate the environmental impact of fine chemical products. Such analysis is based on the comparation of the life cycles of products or manufacturing processes from "cradle to grave" (Figure 3.4). The analysis also considers the consumption behavior of end users, recycling policy and disposal options. In this assessment, the environmental impact is described based on six categories:

1) Raw materials consumption;
2) Energy consumption;
3) Land use;
4) Air and water emissions and solid waste;
5) Potential toxicity;
6) Potential risks.

Combining these individual data yields the total environmental impact of a product or process.

3.4.4
Foreseeable Social and Ethical Aspects

For these applications, the boundaries between SB and earlier processes (production of complex substances by microorganisms, genetically modified or not) are unclear. The production (since the early 1980s) of insulin and other therapeutic proteins by genetically modified organisms has been well accepted.

The major issue will be the reasons for the production of the various compounds (Ro *et al.*, 2006; Khalil and Collins, 2010). Fine chemicals might be highly toxic substances produced as weapons! The case of vitamin production is particularly interesting. Apart from specific projects such as rice supplemented with vitamins, the addition of vitamins is often more a selling point than a true biological need.

Production of such compounds by synthetic organisms will be confined to specific buildings, with strict rules of manipulation. The organisms will be specifically designed to be unable to live in a normal environment. All these issues have already been solved years ago.

Production of these chemical compounds will require a highly sophisticated technology to construct the microorganisms, to purify the chemical compounds and to check their properties. It is difficult to imagine how such a production might be easily shared.

Using SB-based techniques, new compounds may be produced with unknown effects and consequences. This requires centers of expertise with associated infrastructural developments. The developed world will benefit and the developing world will lose out, particularly in the competition for land and water used to produce the biomass that underpins a SB manufacturing industry.

3.5 Cellulosomes

3.5.1 Introduction

Cellulosomes are complexes of enzymes with cellulolytic activity. They are extracelluar scaffoding of proteins bound with enzymes produced by bacteria. Cellolosomes can be used to degrade cellulose and hemicellulose, and they are considered to be "one of the nature's most elaborate and highly efficient nanomachines" (Bayer *et al.*, 1998; Fontes and Gilbert, 2010). Cellulosomal components are integrated via highly ordered protein–protein interactions between two specific proteins: cohesins and dockerins. These specific constructions allow the incorporation of cellulases and hemicellulases onto a molecular scaffold. Cellulosomes can be used for a range of SB applications, from clothes whitening to paperwaste treatment or bio-ethanol production from lignocellulosic materials. The first synthetic cellulosomes have already been constructed (Figure 3.5; Mitsuzawa *et al.*, 2009).

Cellulosomes are complexes of enzymes secreted by anaerobic bacteria that function outside the cell. They were discovered first in the cellulolytic bacterium *Clostridium thermocellum* (Bayer, Kenig, and Lamed, 1983). They function as an efficient strategy for the degradation of insoluble polysaccharides such as those from plant biomass. Currently, the hydrolysis of crystalline substrates like cellulose is an unsolved biochemical problem because no single enzyme can perform this function. The R&D approach to date has been to use a chemical pre-treatment to separate the lignin, hemicellulose and cellulose. Typically, these pre-treatments

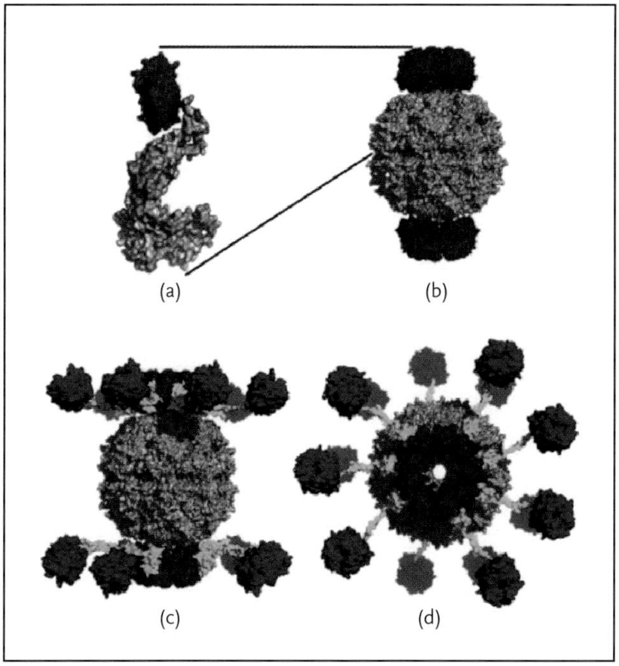

Figure 3.5 Synthetic cellulosome: as recently reported, 18 subunits form this artificial complex (source: Mitsuzawa *et al.*, 2009).

are quite harsh and involve a combination of acid, heat and pressure. Possible alternative methods including microbial and enzymatic methods have only been partially successful. This is because the rate of action on recalcitrant biomass – the resistance of the cellulose molecules in lignocellulosic biomass to enzymatic hydrolysis – remains a technical challenge (McLaren, 2008). An example of the current state of the art in the production of fuel from biomass is provided by Cobalt Technologies. They use beetle-killed lodgepole pine feedstock to produce biobutanol fuel. Mountain pine beetles have apparently infected nearly half of Colorado's five million acres (two million hectares) of pine forest. Millions of acres of lodgepole and ponderosa pines across the western United States and Canada have also been infested, according to Cobalt. Instead of leaving the dead trees as a potential fire hazard, Cobalt said it could use them to make biobutanol. Half of the 2.3 million acres affected in Colorado could produce over two billion gallons (7.5 billion liters) of biobutanol. Cobalt claims this is enough to blend into all the gasoline used in Colorado for six years (Cobalt Technologies, 2010). SB may offer new possibilities for the commercially viable development of the cellulosome complex. Targeted screening for new bacterial strains or directed strain development of existing cellulolytic *Clostridia* species could identify an efficient technological solution for directly converting biomass into fuel substrates. Many technical challenges remain, notably a greater understanding of the bacterial cellulosome mechanism of action (Doi *et al.*, 2003). Specific knowledge of multi-enzymatic complexes on solid surfaces is also necessary to rationally design an optimal enzymatic hydrolysis system for biomass saccharification. The extracellular assembly of the cellulosome and the individual components need to be characterized. The synergism between the individual components and the carbohydrate binding molecules needs to be understood. No artificial formulation of combined cellulases that works effectively at low concentrations has been accomplished. *C. thermocellum* is a cellulosome-producing anaerobic bacterium frequently used as a research model (Bayer, Setter, and Lamed, 1985). It is the most thermophilic among the cellulosome-forming bacteria. It is known for the unique arrangement of cellulosomal genes in its genome and the subunit composition of its cellulosome. The efficiency of its cellulosome makes it a good candidate for commercial bioconversion. Cellulosome technology is not commercially available, and commercializing the process is years to a decade away.

3.5.2
Economic Potential

The importance of SB-tailored cellulose digestion lies in producing biofuels from biomass. The framework has been inspired by the search for sustainable energy substrates as an alternative to fossil fuels. Additionally, SB-based applications also provide tools needed to redesign existing processes. This is particularly valid in the above case because bacteria do not efficiently degrade cellulose.

Cellulosomes have a strong market potential and they should become available commercially. This is due to the worldwide interest in alternative fuels and to the abundance of substrate (feedstock) for SB-based manufacturing processes. Cellulose is one of the most common renewable organic compounds on Earth. About 33% of plant mass is composed by cellulose. The renewed interest of cellulosomes is their ability to address biomass recalcitrance, which is the major bottleneck in the commercialization of cellulose–ethanol production technology. The use of cellulosomes is one of the hot spots in research that aims to address biomass recalcitrance. Using SB techniques have the potential to lower costs and enhance effeciency in mass production. Nonetheless, the technology is not well understood and remains in the research and development stage. Important insights are needed before the technology becomes commercially viable. The urgent demand for the efficient conversion of biomass into biofuels could promote this nascent technology. Currently, the European Union has a Biofuels Directive that sets an indicative (non-binding) biofuels target of 5.75% by energy content by 2010. Most of Europe's transportation fuel demand is for diesel fuel, putting a greater emphasis on biodiesel versus ethanol as the primary biofuel (ethanol, used in the US, is only blended with gasoline). At present, lignocellulose is being considered as a major future feedstock resource for the worldwide production of biofuel from biomass. The United States potential for total biomass (lignocellulose from crop stover consists of the leaves and stalks of maize, forestry, dedicated crops, plus existing crop grains and oils) could be in the order of 1.3 billion tonnes (Perlack *et al.*, 2005).

Lignocellulose has very high potential and, at theoretical conversion rates of 85 gallons ethanol/tonne, could provide biofuels for around 65% of the transport fuel market. The handling of the necessary volumes of biomass and the technical operation of the conversion are not commercially viable today. Nonetheless, a significant amount of research (more than US$ 500 million/year) is being funded to achieve the breakthroughs required to move lignocellulose forward as a viable feedstock. The target is to resolve commercialization barriers in several areas. The main competition in the development of this technology comes from the United States. A case in point is their strong investment program in SB manufacturing infrastructure, for example the US$ 500 million Energy Biosciences Institute (led by BP and the US Department of Energy). Here, SB will have a significant role to play.

There is a significant risk in developing competing technologies for this profitable area because alternatives could become commercially available before the cellulosome technology is ready for market. For example, a parallel approach to utilizing lignocellulose biomass can be applied in gasification systems to produce syngas (Kavalov and Peteves, 2005). Syngas can be used to make electricity by driving a turbine. It can also be a source of hydrogen, or biohydrogen in this case, or can be used as the substrate for other synthetic fuels. Fischer–Tropsch (FT) fuel is one such type. It can be produced from a variety of nonpetroleum feed stocks such as natural gas, coal, petroleum coke, or even biomass and various domestic wastes. If the cellulose biomass deconstruction route does not develop into a sustainable approach, then gasification of lignocellulose may be used as an alternative way of producing cleaner fuels in the short term, while such approach does not currently use cellulosome technology.

3.5.3
Environmental Impact

The paper industry is extremely polluting. Much of the pollution derives from the lack of treatment of lignin-rich effluents. Moreover, no chemicals are available to "solubilize" ligno-cellulosic material. This calls for an integrated view of the cellulose demand upstream of SB processes. The assumption here is that the role of cellulose is to provide a major source of solid substrate to grow cell factories. At the same time, cellulose is insoluble under normal conditions thus probably difficult to use by microbes. Another reason of the low yield of cellulose is that it is combined with lignin. This presumably reflects a very strong selection pressure that permitted (immobile) plants to resist being systematically eaten by microbes.

It therefore remains uncertain whether SB, *per se*, is the best way to degrade cellulose (in fact, the best approach would involve degrading ligno-cellulosic material). Certain proprietary ideas linked to cellulose degradation involve physico-chemical processes rather than biological approaches. It is therefore difficult to foresee the type of environmental impact that might occur.

Note here that lignin is much more important in terms of mass and availability than pure cellulose. Its degradation might create a totally unexpected one since the production of aromatics from lignin is a strategy developed by living organisms to support stable structures. Creating, or using SB technology to select organisms that degrade lignin efficiently might have unwanted consequences if they could multiply in the environment. This consideration is valid for cellulose as well.

During the billions of years of evolution, organisms have developed only very limited ways to degrade cellulose and lignin efficiently. This suggests that there is no easy solution and that the better than natural approaches developed by SB technologies might be rewarding.

3.5.4
Foreseeable Social and Ethical Aspects

Synthetic cellulosomes might dramatically increase the efficiency of cellulose and hemicellulose degradation. The societal impact of their full implementation as a tool for waste treatment and biofuel production would be high, as described above for these two issues.

Could the application change social interactions? The successful application of SB to cellulosomes would, once again, create a new business with the corresponding social classes (producers, traders, sellers, consumers, etc.). The nature of the social interactions is not expected to differ from that of any other new and successful energy business.

Ethical issues. Ethical issues of applying SB to improve cellulosomes for bioethanol production include the same ethical issues as those of bioethanol (see above). For both biofuel production and solid waste management technologies from cellulosomes, intellectual property rights might be particularly important. This would reflect the complexity of the constructions and the expensive research in this field. The negative perception of engineered organisms would be augmented by the "nanomachine" nature of engineered cellulosomes. This might increase the opposition to their use, at least in open environments such as windrow composting.

Justice of distribution. As mentioned above, the complexity of the research and the involvement of intellectual properties mean that a few companies might ultimately control synthetic cellulosomes. Similarly to cellulolytic enzymes, monopolies in controlling cellulosomes might arise. Accordingly, the price and the availability of these complexes would therefore be limited to developing countries.

3.6
Recommendations for Biomaterials

SB will significantly affect the biomaterials market, particularly in the areas of fine chemicals and bioplastics. A toolbox of products serving as biodegradable materials is recommended. The **bulk chemicals industry** will also be significantly affected by the SB-based technology. Putting this into practice will take time and therefore the environmental impacts will be slower. When the new practices are adopted, however, changes will last longer and take place on a much larger scale. In the **fine chemicals industry** the incentives for investment relate to the economic potential of the end product (in contrast to bulk chemical manufacturing). The payoffs could have the major environmental benefits, although these may be largely limited to more efficient use of energy (the core manufacturing practice would still rely on petrochemicals). SB-based techniques may also have some effect on avoiding recalcitrant molecules in the production process. Investment in SB processes in the bulk chemicals sector is likely to have a positive overall impact on plant operations. Examples include the use of biodiesel and less overall chemical waste. Due to the scale of these processes, small changes may have significant positive environmental benefits. For both fine and bulk chemical production, we recommend the deployment of the "chemical bulding block system" as designed (or similar to) the US DoE Figure 3.1.

The field of biopolymers and **bioplastics** urgently needs revisiting in terms of its current labeling for recycling purposes. Categorization of the various products is extremely complex, with negative economic consequences (bioplastics are not necessarily biodegradable). We recommend applying a method clearly outlining which bioplastics need recycling and which can be composted. This should be in place before large-scale commericalisation of SB for production of bioplastics take place. An urgent need to develop completely biodegradable plastics exists. It would benefit from focused SB research and development. There is also a pressing need for **high-performance structural bioplastics** for manufacturing coupled with completely biodegradable additives. Both of these significant growth areas in the bioplastics industry could be greatly improved by SB-based research.

Investment is particularly needed in research and development for new methods and products that will expand and develop tools and manufacturing processes with reduced environmental impact (compared with current manufacturing approaches). Adequate biosafety guidelines need to be established for large-scale manufacturing units. **Cellulosomes** (complex molecules that degrade hemi- and lignocellulosic material) have high economic potential for biofuels, paper and waste processing. SB has the potential to design more efficient and completely new cellulosome complexes to make new, efficient cellulose-digesting proteins. Open sourcing of the cellulosome technology is recommended to address the justice of distribution issues involved in the technology.

References

3. BIOMATERIALS

3.1. BIOMATERIALS IN GENERAL

Carothers, J.M., Goler, J.A., and Keasling, J.D. (2009) Chemical synthesis using synthetic biology. *Curr. Opin. Biotechnol.*, **20**, 498–503.

Columbia Forest Products (2010) Pure bond: formaldehyde-free plywood technology, http://www.columbiaforestproducts.com/PureBond.aspx (accessed 28 May 2010).

Enderle, J., Blanchard, S., and Bronzino, J. (1999) *Introduction to Biomedical Engineering*, Academic Press, New York, p. 538.

Hills, L. (2008) Biomimicry and the built environment, http://www.arabianbusiness.com/538939-biomimicry-and-the-built-environment (accessed 28 May 2010).

Keasling, J.D. (2007) Synthetic biology for synthetic chemistry. *ACS Chem. Biol.*, **3**, 64–76.

London Metal Exchange (2010) Price review: plastics, http://www.lme.com/july05-article4.asp (accessed 28 May 2010).

Markets&Markets Report (2009) Global biomaterial market 2009–2014, http://www.the-infoshop.com/report/mama99910-glob-bio-mtrl.html (accessed 28 May 2010).

USDOE (2004) Top value added chemicals from biomass. Volume I: results of screening for potential candidates from sugars and synthesis gas, http://www1.eere.energy.gov/biomass/pdfs/35523.pdf (accessed 28 May 2010).

3.2. BIOPOLYMERS/PLASTICS

Aldor, I.S., and Keasling, J.D. (2003) Process design for microbial plastic factories: metabolic engineering of polyhydroxyalkanoates. *Curr. Opin. Biotechnol.*, **14** (5), 475–483.

Anastas, P., and Warner, J. (1998) *Green Chemistry: Theory and Practice*, Oxford University Press, London.

ENDS Report (2002) Co-op launches degradable carrier bags – amid inflated green claims. **332**, 35–36.

ENDS Report (2004) Global interest grows in biopolymers. **315**, 29–32.

ENDS Report (2008) Government plans joined-up food strategy. **402**, 47–48.

ENDS Report (2009) When will bioplastics finally grow up? **410**, 40–44.

ENDS Report (2010) Label for home-compostable packaging ready. **422**, 20.

Ethanol Business (2007) Brazil braskem: to build unit for polyethylene from ethanol, http://ethanol-business.com/2007/10/30/brazil-braskem-to-build-unit-for-polyethylene-from-ethanol/ (accessed 28 May 2010).

European Bioplastics Conference (2007) Biopolymers: market potential and research challenges. Interplast, http://www.inter-plast.info/LinkClick.aspx?fileticket=5coYZ5aamUw%3D&tabid=74&language=it-IT (accessed 28 May 2010).

Innovia Films (2010) NE. http://www.innoviafilms.com/products/brand/natureflex1/ne30 (accessed 28 May 2010).

Jones, R. (2009) Economics, sustainability, and public perception of biopolymers. *Soc. Plast. Eng.*, doi: 10.1002/spepro.00060.

Lee, S.Y., and Lee, Y. (2003) Metabolic engineering of *Escherichia coli* for production of enantiomerically pure (R)-(–)-hydroxycarboxylic acids. *Appl. Environ. Microbiol.*, **69** (6), 3421–3426.

London Metal Exchange (2010) Price review: plastics, http://www.lme.com/july05-article4.asp (accessed 28 May 2010).

Michael, D. (2003) Biopolymers from crops: their potential to improve the environment. Australian Society of Agronomy. http://www.regional.org.au/au/asa/2003/c/11/michael.htm (accessed 28 May 2010).

Mirel (2010) Discover Mirel, http://www.mirelplastics.com/discover/default.aspx?ID=1516 (accessed 28 May 2010).

Natureworks LLC (2010) Ingenious materials, http://www.natureworksllc.com/ (accessed 28 May 2010).

Novamont (2010) Mater-Bi® is applied in a variety of industrial sectors, http://www.materbi.com/default.asp?id=547 (accessed 28 May 2010).

Plastics2020challenge (2010) Do bioplastics represent the solution for the future?

http://plastics2020challenge.com/2009/07/06/do-bio-plastics-represent-the-solution-for-the-future/ (accessed 28 May 2010).
Rao, K. (2010) The current state of biopolymers and their potential future, http://www.omnexus.com/resources/editorials.aspx?id=25041 (accessed 28 May 2010).
Rhu, D.H., Lee, W.H., Kim, J.Y., and Choi, E. (2003) Polyhydroxyalkanoate (PHA) production from waste. *Water Sci. Technol.*, **48** (8), 221–228.
Solanyl (2008) What is Solanyl Bioplastic? http://www.biopolymers.nl/en/bioplastic/ (accessed 28 May 2010).
Solvay Plastics (2010) Who are we? http://www.solvayplastics.com/services/strategy/solvayindupa/0,76941-2-0,00.htm (accessed 28 May 2010).
Synbra (2009) BioFoam®: a new biodegradable EPS, http://www.synbra.com/en/38/191/55/nieuws/biofoam174_a_new_biodegradable_eps.aspx (accessed 28 May 2010).
Treehugger (2007) Ecoflex® compostable plastic packaging materials by BASF, http://www.treehugger.com/files/2007/11/ecoflex_compost_1.php (accessed 28 May 2010).
Widmaier, D.M., Tullman-Ercek, D., Mirsky, E.A., Hill, R., Govindarajan, S., Minshull, J., and Voigt, C.A. (2009) Engineering the salmonella type III secretion system to export spider silk monomers. *Mol. Syst. Biol.*, **5**, 309–317.

3.3. BULK CHEMICAL PRODUCTION

Biopact (2007) Metabolic explorer partners with IFP to develop propanediol from biodiesel byproduct glycerine, http://news.mongabay.com/bioenergy/2007/09/metabolic-explorer-partners-with-ifp-to.html (accessed 28 May 2010).
Cargill (2008) Cargill and Novozymes to enable production of acrylic acid via 3HPA from renewable raw materials, http://www.cargill.com/news-center/news-releases/2008/NA3007665.jsp (accessed 28 May 2010).
Carothers, J.M., Goler, J.A., and Keasling, J.D. (2009) Chemical synthesis using synthetic biology. *Curr. Opin. Biotechnol.*, **20** (4), 498–503.
Dornburg, V., Hermann, B.G., and Patel, M.K. (2008) Scenario projections for future market potentials of biobased bulk chemicals. *Environ. Sci. Technol.*, **42** (7), 2261–2267. http://pubs.acs.org/doi/abs/10.1021/es0709167 (accessed 28 May 2010).
Dow (2010) Dow advanced materials CEO says clean and sustainable technologies poised to address over $20 billion in market opportunities, http://news.dow.com/dow_news/corporate/2010/20100317a.htm (accessed 28 May 2010).
Green Chemicals (2009) Jessica Biel and stars make a splash at Dow live earth run for water launch event, http://www.icis.com/blogs/green-chemicals/2009/10/jessica-biel-and-stars-come-ou.html (accessed 28 May 2010).
Hermann, B.G., and Patel, M.K. (2007) Today's and tomorrow's bio-based bulk chemicals from white biotechnology: a techno-economic analysis. *Appl. Biochem. Biotechnol.*, **136** (3), 361–388.
Hermann, B.G., Blok, K., and Patel, M.K. (2007) Producing bio-based bulk chemicals using industrial biotechnology saves energy and combats climate change. *Environ. Sci. Technol.*, **41** (22), 7915–7921.
ICIS Green Chemicals (2010) Wacker licenses bio-acetic acid process, http://www.icis.com/blogs/green-chemicals/2010/04/wacker-licenses-bio-acetic-aci.html (accessed 28 May 2010).
Jeremijenko, N. (2004) Amateurity and biotechnology, http://www.locusplus.org.uk/NJ01.pdf (accessed 28 May 2010).
Keasling, J.D. (2007) Synthetic biology for synthetic chemistry. *ACS Chem. Biol.*, **3**, 64–76.
Li, S., Tuan, V.A., Falconer, J.L., and Noble, R.D. (2001) Separation of 1,3-propanediol from solutions by pervaporation using a ZSM-5 zeolite membrane. *J. Memb. Sci.*, **191**, 53–59.
NEST Report (2005) Synthetic biology applying engineering to biology: report of a NEST high-level expert group, http://ftp.cordis.europa.eu/pub/nest/docs/syntheticbiology_b5_eur21796_en.pdf (accessed 28 May 2010).
New Scientist (2010) Better living through green chemistry, http://www.newscientist.com/article/mg20527511.

500-better-living-through-green-chemistry.html (accessed 28 May 2010).
Patel, M., Crank, M., Dornburg, V., Hermann, B., Roes, L., Hüsing, B., Overbeek, L., Terragni, F., and Recchia, E. (2006) Medium and longterm opportunities and risks of the biotechnological production from renewable resources, http://www.chem.uu.nl/brew/BREW_Final_Report_September_2006.pdf (accessed 28 May 2010).
Piribo (2010) Global bulk drugs industry report, http://www.piribo.com/publications/general_industry/global_bulk_drugs_industry_report.html (accessed 28 May 2010).
Tanewiski, M. (2006) Sustainable chemical technologies. Development trends and tools. *Chem. Eng. Technol.*, **29** (12), 1397–1403.

3.4. FINE CHEMICAL PRODUCTION

AllBusiness (2010) Verdezyne produces adipic acid biologically, http://www.allbusiness.com/science-technology/biology-biodiversity/13886056-1.html (accessed 28 May 2010).
Amgen (2005) Amgen Manufacturing, http://www.amgen.no/pdfs/Fact_Sheet_Manufacturing.pdf (accessed 28 May 2010).
BASF (2005) Asessing biobased materials and processes with the eco-efficiency analysis and SEEbalance® of BASF, http://www.basf.com/group/corporate/de/function/conversions:/publish/content/sustainability/eco-efficiency-analysis/images/Biobased_materials_Gent_200905_Saling.pdf (accessed 28 May 2010).
BASF (2010) What is eco-efficiency, http://www.basf.com/group/corporate/en/sustainability/eco-efficiency-analysis/what-is (accessed 28 May 2010).
Bremus, C., Herrmann, U., Bringer-Meyer, S., and Sahm, H. (2006) The use of microorganisms in l-ascorbic acid production. *J. Biotechnol.*, **124** (1), 196–205.
C&EN (2008) Optimism prevails in fine drugs market, http://pubs.acs.org/cen/coverstory/86/8604cover3.html (accessed 28 May 2010).
Carpenter, K.J. (2001) Chemical reaction engineering aspects of fine chemicals manufacture. *Chem. Eng. Sci.*, **56** (2), 305–322.
Deinove (2009) Deinococcus sp, http://www.deinove.com/en/research-and-development/deinococcus-sp.html (accessed 28 May 2010).
DiMasi, J.A., Hansen, R.W., and Grabowski, H.G. (2003) The price of innovation: new estimates of drug development costs. *J. Health Econ.*, **22**, 151–185.
EuropaBio Report (2003) White biotechnology: gateway to a more sustainable future, http://www.mckinsey.com/clientservice/chemicals/pdf/biovision_booklet_final.pdf (accessed 28 May 2010).
Global Business Coalition (2008) Sanofi-Aventis, OneWorld, Amyris partner to develop artemisinin to treat malaria, http://gbcimpact.org/itcs_node/0/0/news/603 (accessed 28 May 2010).
Global Information Inc. (2004) The U.S. market for nutritional supplements: vitamins, minerals and dietary supplements, http://www.the-infoshop.com/report/pf23739_us_supplements.html (accessed 28 May 2010).
Hale, V., Keasling, J.D., Renninger, N., and Diagana, T.T. (2007) Microbially derived artemisinin: a biotechnology solution to the global problem of access to affordable antimalarial drugs. *Am. J. Trop. Med. Hyg.*, **77** (6), 198–202.
Khalil, A.S., and Collins, J.J. (2010) Synthetic biology: applications come of age. *Nat. Rev. Genet.*, **11**, 367–379.
Riva, S. (2006) Laccases: blue enzymes for green chemistry. *Trends Biotechnol.*, **24** (5), 40–43.
Ro, D.K., Paradise, E.M., Ouellet, M., Fisher, K.J., Newman, K.L., Ndungu, J.M., Ho, K.A., Eachus, R.A., Ham, T.S., Kirby, J., Chang, M.C.Y., Withers, S.T., Shiba, Y., Sarpong, R., and Keasling, J.D. (2006) Production of the antimalarial drug precursor artemisinic acid in engineered yeast. *Nature*, **440**, 940–943.
Roumeliotis, G. (2006) North American fine chemicals market to reach $28.62 bn in

2011. In-Pharma. http://www.in-pharmatechnologist.com/Materials-Formulation/North-American-fine-chemicals-market-to-reach-28.62bn-in-2011 (accessed 28 May 2010).

OECD (2001) The application of biotechnonoly to industrial sustainability, http://www.oecdbookshop.org/oecd/display.asp?lang=EN&sf1=identifiers&st1=932001061p1 (accessed 28 May 2010).

Sabot, O.J., Mwita, A., Cohen, J.M., Ipuge, Y., Gordon, M., Bishop, D., Odhiambo, M., Ward, L., and Goodman, C. (2009) Piloting the global subsidy: the impact of subsidized artemisinin-based combination therapies distributed through private drug shops in rural Tanzania. *PLoS ONE*, **2009**, 40–43. http://www.plosone.org/article/info:doi%2F10.1371%2Fjournal.pone.0006857 (accessed 28 May 2010).

Shafaati, A. (2007) Chiral drugs: current status of the industry and the market. *Iran. J. Pharm. Res.*, **6** (2), 73–74.

U.S. Department of Energy (1999) New biocatalysts: essential tools for a sustainable 21st century chemical industry, http://www.dtic.mil/cgi-bin/GetTRDoc?AD=ADA436521&Location=U2&doc=GetTRDoc.pdf (accessed 28 May 2010).

3.5. CELLULOSOMES

Bayer, E.A., Kenig, R., and Lamed, R. (1983) Adherence of clostridium thermocellum to cellulose. *J. Bacteriol.*, **156** (2), 818–827.

Bayer, E.A., Setter, E., and Lamed, R. (1985) Organization and distribution of the cellulosome in clostridium thermocellum. *J. Bacteriol.*, **163**, 552–559.

Bayer, E.A., Shimon, L.J.W., Shoham, Y., and Lamed, R. (1998) Cellulosomes: structure and ultrastructure. *J. Struct. Biol.*, **124**, 221–234.

Cobalt Technologies (2010) Biobutanol and beyond, www.cobalttech.com (accessed 28 May 2010).

Doi, R.H., Kosugi, A., Murashima, K., Tamaru, Y., and Han, S.O. (2003) Cellulosomes from mesophilic bacteria. *J. Bacteriol.*, **185** (20), 5907–5914.

Fontes, C.M., and Gilbert, H.J. (2010) Cellulosomes: highly efficient nanomachines designed to deconstruct plant cell wall complex carbohydrates. *Annu. Rev. Biochem.*, **79**, 40–43.

Kavalov, B., and Peteves, S.D. (2005) Status and perspectives of biomass-to-liquid fuels in the European Union. Report EUR 21745 EN.

McLaren, J.S. (2008) The economic realities, sustainable opportunities, and technical promises of biofuels. *AgBioForum*, **11** (1), 8–20. www.agbioforum.org (accessed 28 May 2010).

Mitsuzawa, S., Kagawa, H., Li, Y., Chan, S.L., Paavola, C.D., and Trent, J.D. (2009) The rosettazyme: a synthetic cellulosome. *J. Biotechnol.*, **143** (2), 139–144.

Perlack, R.D., Wright, L.L., Turhollow, A.F., Graham, R.L., Stokes, B.J., and Erbach, D.C. (2005) Biomass as feedstock for a bioenergy and bioproducts industry: the technical feasibility of a billion-ton annual supply. DOE/GO-102005-2135.

Further Reading

Market Research.com (2010) The future of blood disorders therapeutics, analysis and market forecasts to 2016: better and more cost-effective treatment options create opportunities, http://www.marketresearch.com/product/display.asp?productid=2623011 (accessed 28 May 2010).

4
Other Developments in Synthetic Biology

Rachel Armstrong, Markus Schmidt, and Mark Bedau

4.1
Protocells

4.1.1
Introduction

Protocells are currently models of artificial cells that have some properties of living systems but are not yet fully alive. They have not yet been developed to full "artificial cell" status, although it is expected to happen within the next 5–10 years. Protocells embody a different kind of approach to the practice of synthetic biology: they are designed using a "bottom up" approach to assembly using basic chemistry. This differs from the "top down" perspective used by conventional synthetic biology, which "hacks" existing biological systems, using DNA, to reassign function or add new functionality. The latter "hacks" existing biological systems, using DNA, to reassign function or add new functionality. Currently, protocells serve as laboratory research models; no commercial applications have yet been explored (Figure 4.1).

The properties of existing protocell models employ the capabilities of self-assembling chemical systems that are distributed spatially and temporally by virtue of their life-like properties such as mobility, chemotaxis and agent/agent interactions. They are embodied in a container and are able to interact with a local environmental context by virtue of a combination of the properties (e.g., oil) and additional chemistry (e.g., anhydride) that is distributed spatially and temporally by the life-like properties of the oil droplets. Ongoing research is exploring potential applications for protocell systems in situations that require embodied complex solutions or environmental responsiveness. Eventually, fully artificial cells will participate in sustainable manufacturing and environmental practices in a number of ways.

Protocells have been proposed as delivery systems for "environmental pharmaceuticals", for example, the development of smart "paints" or surface coatings that can fix carbon dioxide into inorganic carbonate. Protocell systems are capable of multiple functions, for example, protocell "paints" not only will act as an accretion

Synthetic Biology: Industrial and Environmental Applications, First Edition. Edited by Markus Schmidt.

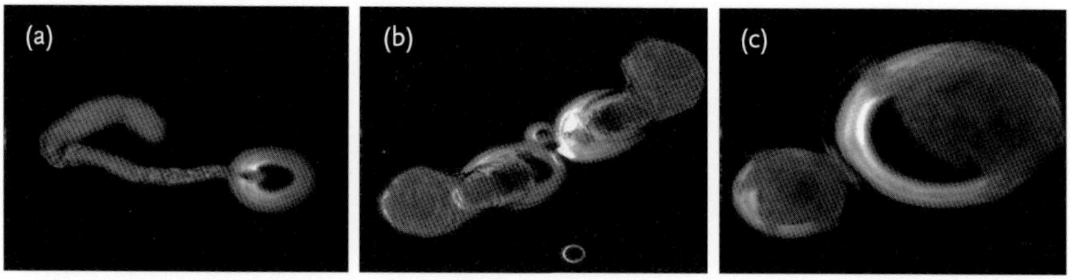

Figure 4.1 Protocells exhibit interactive and possibly even "social" behavior, appearing to share a chemical "language" which appears as "fingers" (a) produce material at the interface of adjacent agents (b) and exhibit life-like properties including movement and the shedding of skins (c). Photographs by Rachel Armstrong (Armstrong and Spiller, 2009).

technology to "fix" carbon dioxide from the surroundings but also will improve the thermal insulation qualities of existing buildings by creating a thicker wall as a consequence of producing a precipitate of inorganic carbonate and will improve the energy efficiency of the building.

Protocells have also been proposed as complex chemical agents that can perform basic conversion reactions such as cleaning up environmental waste toxins like cyanide and converting them into harmless thiocyanide that can be absorbed into the natural ecological system.

Protocellls may also have been integrated into existing systems so that they can metabolically participate in a synthetic ecology and assist traditional synthetic biology in achieving some of its goals. One example is helping extremophile bacteria to perform under extreme conditions by providing a slow release system of inorganic nutrients (The Green Optimistic, 2010).

Protocell technology may also be combined with solar fuel producing chemistry to provide alternative methods of creating biofuels from sunlight and carbon dioxide (so-called carbon capture and recycling; CCR technology).

Currently, protocells are a research technology. They are produced in very limited quantities by hand and cannot replicate because they lack DNA. Protocells have a life span of minutes and this needs to be extended before they can become commercially exploitable. Protocell research tends to emphasize the biochemical synergies and other unpredictable emergent properties found in even the simplest forms of life (Rasmussen *et al.*, 2008) and suggests that these artificial cells will provide a new platform with the potential to coordinate these processes. These synergies and emergent properties mean that protocell engineering is not as predictable or reliable as programming a computer. Developments in protocell design approaches (using computer or chemical modeling) will be needed to create a platform where protocells with specific functions can be (relatively) reliably "evolved". Protocells have technological limitations: they cannot operate in dry environments, and they currently need either oil or water as a medium to function "metabolically" or chemically. Studies on protocell interactions with existing organisms have not been carried out, and their environmental toxicity is unknown. Additionally, their risk to "natural" systems therefore cannot be evaluated, since studies on protocell interactions with existing organisms have not been carried out, so their environmental toxicity is unknown. However, protocells are not self-sustaining, and the current models are not fully alive since they cannot self-replicate. Ongoing research is aimed at addressing the key technical challenge in the production of protocell technology, which is to create an adequate infrastructure for their growth, division and replication (through a heritable mechanism of information). Although protocells are currently an immature research technology, they will be increasingly driven towards commercial applications by private enterprise. The first commercial carbon-fixing building coatings could be available within 5–10 years. In these systems the protocells will not be fully artificial cells but will function over a period of months to produce an accretion of carbonate on the surface to which they are applied. Further applications of the "paint" would need to be applied in order to keep the growing surface "active". A number of paint companies are currently considering protocells as "smart", environmentally friendly agents that could be used in emulsion-based exterior paints (New Scientist, 2010). Smart surface coatings that are self-regenerating and rely on self-regenerating protocells are probably another 20 years away. The recently announced Carbon Prize includes protocell technology in its portfolio as a potential CCR platform (Prize Capital LLC, 2010).

4.1.2 Economic Potential

Protocells are currently not commercially available and therefore have no existing market. However, there is early commercial interest in protocell applications as they can be envisioned as entering either the fine chemicals market or having environmental applications (decontamination and carbon capture) depending on the proposed context of their use. Depending on the proposed context of their use, however, they can be envisioned as entering either the fine chemicals market or having environmental applications (decontamination and carbon capture). No market is established for protocells.

The current investment and interest in the model research protocell systems lies in the novelty of the technology as a complex chemical technology that offers new problem-solving tools. A major weakness in their development towards commercial applications is that potential products are still at very early stages of research. Nonetheless, the opportunities for model protocells based on self-assembling chemistry are significant. The potential lies in environmental remediation as well as notional carbon capture and recycling, without the controversy associated with genetic modification. The technology is threatened by its very early stage of application and its research and development stage of production.

4.1.3 Environmental Impact

Protocells have the potential to occupy a unique position within an environmental context. This is because they represent a transitional interface between traditional chemistry and biology. The flexibility of biological systems is combined with the durability of a chemical system. The result is a new set of problem-solving tools for dealing with environmental challenges. Moreover, they appear to be compatible with existing biological technologies.

Protocell technology can be thought of as a new platform for the slow release delivery of chemicals into stressed environments. It performs the role of "environmental pharmaceuticals" that restore a system's balance in the event that substances are present in excess (toxins) or are depleted (nutrients). In the built environment, protocell technology can help to recapture carbon emissions and, importantly, recycle these into building materials. This approach is currently carbonate-based, but other architectural materials could conceivably be synthesized. This represents a new generation of "retrofitting" approaches: rather than destroying existing architectures to renovate them, building exteriors are treated with accretion technologies based on protocells. This improves thermal insulation properties, heals and repairs fractures in the exteriors, and possibly even helps develop environmentally sensitive and responsive materials whose growth is influenced by environmental conditions. For example, an abundant supply of carbon dioxide would promote the growth of a building because the supply of raw material for the accretion process is abundant. This growth, however, is rate-limited by the cation concentrations in the applied protocell technology (Armstrong, 2009; Armstrong and Spiller, 2009).

In an architectural context, the existence of self-assembling, smart materials would generate a new way of thinking about the spatial design of cities, use of construction materials, fate of rooftop water and the bioactivity of building facades. Accordingly, the surface of the built environment represents a dynamic interface that enables artificial and natural metabolic processes to interact. This would yield design and environmental outcomes that positively influence the urban environment. Currently, the surface of the built environment is being enlisted to help reduce fossil fuel consumption. The simple concept: paint the reflective surfaces of buildings white (proposed by Stephen Chu). This simple change alone would radiate enough heat to reduce greenhouse gas emissions by 44 billion tonnes (calculated in collaboration with Art Rosenfeld from the Californian Energy Commission, using data based on the reflective surfaces available in the 10 largest cities; Barringer, 2009).

Additionally, the application of accretion materials in an architectural context simply differs from the current practices, which are based on industrial technologies with high fossil fuel-consuming indices. Repeated applications of accretion materials could be used in ecologically sensitive sites. One example would be the sustainable reclamation of Venice using protocell technology to grow an artificial reef underneath the city, which has a tempestuous relationship with the sea (Devlin, 2009).

The sustainability of the technology depends on the chemical formula of the agents, which are based on oil and water. Other potential substrates for future research as delivery matrixes (containers) include complex carbohydrates. These molecules are biodegradable and sources of food for organisms. Therefore the toxic potential of these systems can be determined and their context designed a priori from the initial ingredients. Environmental risks posed by protocells may come from pollution due to the oil residues and chemical deposits they produce. These componds may spoil sites in which the "environmental pharmaceuticals" are applied.

The combination of protocell technology with solar fuel chemistry is at an early stage of development. This combination may prove to have several advantages over current biofuel production methods: the chemistry does not respire in the absence of light, and it can be applied to surfaces like solar panels (making land in cities available to create urban biomass). Finally, the local production of biofuels would greatly reduce the fossil fuel-burning transportation methods needed to take the product to users.

4.1.4
Foreseeable Social and Ethical Aspects

Due to the fact that protocells do not yet exist as "artificial life", no special measures or institutions are in need today to regulate them. The model systems being used to develop the technology have environmentally useful applications. This calls for thinking today about preparing for our future with protocells. The ethical, social and regulatory issues that the first fully artificial cell will generate need to be carefully thought through, particularly given the recent technological success of J Craig Venter's Systems Genomics group. They created the first synthetic biology organism with an entirely artificial and operational gene sequence and termed it *Mycoplasma mycoides* JCV1-syn1.0 (Gibson *et al.*, 2010). Bedau *et al.* (2010) recommend six checkpoints for the social and ethical development of protocells. These are: (i) systemic inclusion of social and ethical considerations into basic research and education programs, (ii) technical checkpoints to assess whether the creation of "artificial life" in the laboratory has been successfully achieved, (iii) creating the first "artificial cell" (which carries the most ethical and social importance), (iv) environmental impact assessment, (v) release into the environment and (vi) assessment of toxicity. Although these technologies are still immature, they have some potential in offering new ways of addressing climate change as an additional strategy to energy conservation. Such protocell systems will probably have the potential to be self-sustaining within the next 10 years. They will become increasingly important in combating the chemical changes underpinning global warming through surface applications in the built environment, where most emissions are produced.

Social interaction will change, in that some of the current "burden" of environmentally responsible social practice (e.g., selection of biofuels, recycling, or even choosing not to burn candles on a birthday cake) could be taken over by technology. Potentially, the design of surfaces at the source of pollution by chemical and combustion waste could at least in part involve protocell-based systems. This generally saves time and resources in societies where recycling and carbon consciousness are a significant factor in daily choices, routines and publicly funded services. The creation of life in the laboratory is a profoundly important ethical issue to address.

Widespread inequalities in access based on economics, health care systems and geography are to be expected. This inequity could be addressed by evaluating the potential for "technology transfer" at an early stage of protocell technology. This would facilitate the local uptake and development of protocell-based carbon dioxide fixing technologies in those areas responsible for the most carbon dioxide emissions (India, China). Open sourcing of protocell technology may help address some of these inequalities in the justice of distribution.

4.2 Xenobiology

4.2.1 Introduction

The concept of *modularity* has been introduced since the industrialization. It has experienced previously unimaginable levels of innovation and growth. Modularity means building complex products from smaller subsystems that can be designed independently yet function together as a whole. Modularity freed designers to experiment with different approaches, as long as they obeyed the established design rules. One of the key requirements of modularity, however, is orthogonality. The term orthogonality stems from the Greek orthos, "straight", and gonia, "angle". The term was originally used to describe the mathematical situation in which two vectors are perpendicular, that is, form a right angle. Changes in the magnitude of one vector do not affect the magnitude of the other vector. In engineering, orthogonality is a system design property facilitating the feasibility and simplicity of complex designs. Orthogonality guarantees that modifying one component of a system does not propagate side effects to other components of the system. With the clear benefit of orthogonality in complex systems in mind, synthetic biologists are now attempting to apply these engineering principles to biology. While engineers have been quite successful applying the principles of orthogonality to the non-living world, biologists still have to overcome major challenges. This is because natural life forms hardly exhibit a true orthogonal design. The efforts undertaken by synthetic biologists to construct orthogonal biological systems are twofold, focusing either on the metabolism or on the biochemical building blocks. Xenobiology is the modification and exchange of some of the elementary biochemical building blocks of life.

The focus has been directed to the research on alternative biomolecules that can sustain living processes (Schmidt, 2010). Areas of research include chemically modifying amino acids, proteins or DNA, yielding:

- Enlarged genetic alphabets (Yang *et al.*, 2006; Yang *et al.*, 2007; Leconte *et al.*, 2008);
- Xeno nucleic acids (Herdewijn and Marliere, 2009);
- Mirror life (biochemicals with opposite chirality; Carr and Church, 2009);
- Quadruplet codons (Anderson *et al.*, 2004; Neumann *et al.*, 2010);

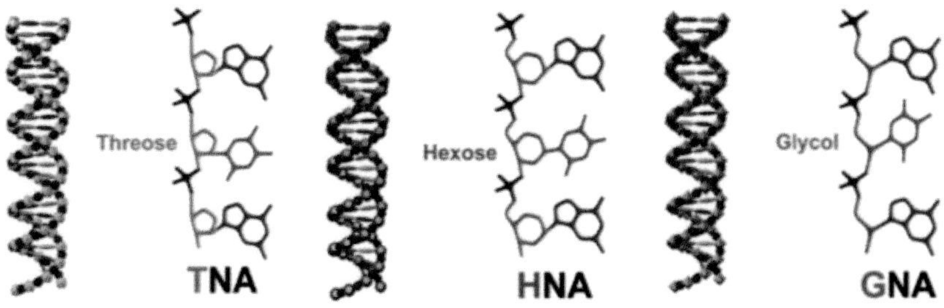

Figure 4.2 Xeno nucleic acids are alternatives to DNA as information-storing biopolymers (source: Schmidt, 2010).

- Reassigned codons (Chin, 2006);
- Never-born proteins with canonical and non-canoncial amino acid combinations that do not exist in nature (Budisa, 2004; Luisi, Chiarabelli, and Stano, 2006).

Although many research papers have been published, xenobiology is still in its infancy. No organism based on any of the xeno systems described here exists anywhere in the world. Further progess could, however, bring an unprecedented increase in chemical biodiversity, with a wide range of possible applications, including environmental biotechnology (Figure 4.2).

4.2.2
Economic Potential

Business opportunities are plentiful despite the early phase of xenobiology and xenobiological applications. Theoretically, xenobiological products could replace almost each natural product or organism. Initially, xenobiology will be used only where a unique advantage over existing technologies can be demonstrated.

One of the few (clinical) applications already on the market are branched DNA assays developed by scientists at Chiron and Bayer Diagnostics. In HIV and hepatitis virus diagnostics, the target RNA molecule to be detected (the analyte) is attached to a series of capture DNA probes. These assays use an expanded genetic alphabet (non-canonical base pairs). Using standard nucleotides created significant noise because non-target DNA present in the biological sample was captured by the probes in the microwell even in the absence of analyte. Incorporating components of the artificial genetic alphabet reduced the noise. The assay now helps manage the care of about 400 000 HIV and hepatitis patients annually, detecting as few as eight molecules of the analyte DNA in a sample (Sismour and Benner, 2005).

Another potential market relates to safe GMOs. DNA synthesis and the enzymatic amplification, assembly and recombination of recombinant genetic sequences for industrial purposes have become routine, that is, transforming ecosystems by disseminating synthetic genetic constructs is now a major topic of debate between the lay public and experts (Gaskell *et al.*, 2004). The risks of "genetic pollution" present a challenge that needs to be anticipated now. This calls for designing and deploying technologies for preventing or restricting genetic crosstalk between natural species and the artificial biodiversity. Because XNA building blocks do not occur in nature, they would have to be synthesized and supplied to cells. These cells should be equipped with an appropriate enzymatic machinery for polymerizing them. Human health and natural ecosystems will be more safely conserved by embodying xenobiological material, that is, in replicons, chemically separate from the support on which natural selection has acted so far. This would create new markets for safe GMOs (Herdewijn and Marliere, 2009).

4.2.3 Environmental Impact

Millions of recombinant DNA experiments have been carried out during the past 35 years, and post-Asilomar biosafety guidelines are in place. Apart from infrequent BSL 3 and 4 laboratory accidents, there is no evidence that genetically modified organisms have wreaked havoc on our planet or are the source of major pandemics. If everything is fine right now, why develop xenobiological systems?

Over the medium and long term it makes sense to design and construct a hardware and software of life that is different from "life as we know it". The first 35 years of genetic engineering were a mere prelude to what is expected in the next 35 years and beyond. The upcoming development in bioengineering will be shaped by the following driving forces:

- Key supporting technologies such as sequencing and DNA (XNA) synthesis will become much cheaper and more powerful, a development similar to Moore's law in electronics (Carlson, 2003).
- The design and construction of large biological systems, instead of just modifying single gene, will improve both the speed and depth of genetic engineering.
- R&D experiments will soon be carried out by robots, both physically and conceptually, further decreasing the costs and increasing the number of experiments.
- More people (and their robots) will be able to conduct those experiments. The *de facto* monopoly of academia and industry will soon be gone, giving rise to a new breed of inspired biohackers and amateur biologists (Schmidt, 2008).
- Converging or "living technologies" will increasingly bring together hardware, software and wetware (Ran, Kaplan, and Shapiro, 2009; Bedau *et al.*, 2010).
- DNA is becoming a molecule of choice also for non-biological applications, for example, as templates for nanotechnology self-assembly systems (Kershner *et al.*, 2009).
- Potential public fear–and subsequent regulatory red taping–of fast, in-depth and ubiquitous engineering of our own genetic (source) code could stifle further developments and opportunities.

- The amount and complexity (or depth) of DNA-based engineering will increase by an order of magnitude over the next decades rather than merely doubling or tripling.

Whatever new or improved *physical containment* mechanisms (in contrast to *semantic containment* mechanisms) will be developed, there is one key problem that cannot be solved: all biotech (and nanobiotech) use the same "software program", namely DNA and RNA that occurs in all naturally evolved and domesticated microbes, plants and animals. Xenobiology could become a fundamental safety device capable of limiting any kind of genetic interaction with the natural world. Xenobiology could create no less than an isolated genetic enclave within the natural world (Marliere, 2009). In this scenario, xeno-organisms would be able to maintain all basic functions of life such as compartimentalization, metabolism, replication, reproduction, environmental interaction, growth and so on.

Key differences between the xeno and the natural world remain. Precisely these differences make the "genetic firewall" so interesting in terms of environmental safety.

The xeno-organisms must not and cannot produce certain essential biochemical building blocks, namely their own nucleotides. These biochemicals will have to be supplied externally, and the XNA building blocks should at least be two synthetic steps away from any natural molecule.

Because natural and xeno-organisms will use a different and very specific set of nucleotide-binding proteins for replication and transcription, gene flow – whether horizontal or via sexual reproduction – cannot occur between the two realms of life. XNA can therefore not transfer to wild-type organisms and be incorporated into their DNA genomes subsequently.

Although the exchange of genetic information is not possible, other types of interaction are. For example, the XNA organisms could produce, sense or dismantle chemical substances under laboratory conditions or in the environment. In theory, xeno-organisms could interact with each other to form their own ecosystem. These ecosystems, however, would be small because all organisms need to be supplied with their essential biochemicals. XNA provides a genetic firewall, but not a biological firewall. That means that XNA organisms might interact with DNA organisms on an ecological level, but never on a genetic level (Schmidt, 2010).

Advances in xenobiology research – the creation of "alien" or "weird" life in the laboratory – will not only contribute to a better understanding of the origin of life, but will definitely expand our capabilities to provide safer biotechnology production tools for human and environmental needs. Future life forms that are orthogonal to natural life forms, such as those based on xeno nucleic acids, could represent the ultimate biosafety tool. The more layers of orthogonality, however, the safer. Orthogonal xenobiological systems that act as genetic firewalls to natural life forms can be achieved by a combination of XNA, use of non-canonical base pairs, non-canonical amino acids, alternative codon assignment, even quadruplet codons, or systems different from the tripartite DNA–RNA–protein architectures.

4.2.4 Foreseeable Social and Ethical Aspect

The advent of recombinant DNA technology in the 1970s so worried scientists that they organized the now famous Asilomar conference in 1975 to discuss the risks of genetic engineering. Although not all recommendations of Asilomar were put into practice, it helped avoid the potential negative consequences of this technology (Berg *et al.*, 1975; Berg, 2008). Discussing societal aspects of xenobiology today calls for a consideration of the following aspects (Schmidt *et al.*, 2009): biosafety, biosecurity, intellectual property rights and governance.

In contrast to these rather tangible aspects, we might also be confronted with rather intangible implications. The history of science shows several changes in our worldviews, altering our folk-based narratives to more scientifically inspired (semi-)rational approaches. In this context, science has inflicted a series of disappointments and disillusions to our folk-based beliefs. These include: (i) the earth is not the center of the universe, (ii) men and apes share the same ancestors, (iii) emotions and thinking are correlated to a neurological substrate. The promoters of these ideas were often attacked by those trying to keep the intellectual *status quo*. Xenobiology could easily trigger the next paradigm change in the way we understand nature and life. Just as the Earth lost its place as the center of the universe, or humans lost their unique status in the animal kingdom, our natural world could lose its unique status as being synonymous with "life". As with all other paradigm changes, however, concepts that better explain the world around us cannot be ignored for long (Schmidt, 2010).

4.3 Recommendations for Protocells and Xenobiology

Protocell technology represents a bottom-up approach to synthetic biology, bridging inorganic and organic processes. This technology enables a better understanding of synthetic biology as a whole to develop new technologies. Although the research is at an early stage, the development of potentially radically novel and significant environmental interventions seems feasible, for example, for the remediation of carbon emissions and for alternative biofuels technology. One strong recommendation is to invest in basic science to underpin and support the research while private investment is ongoing.

Protocell technology has a huge potential to offer radically different tools and methods than previously encountered with synthetic biology-based approaches. This is because of its bottom-up nature and because of its overlaps with basic chemistry. The focus should also be on a toolbox of potential products and on the investigation of issues related to make the technology open source. **Xenobiology** (also known as chemical synthetic biology) is another bottom-up approach to design and construct radically new biological systems with properties not found in nature. By using non-canonical amino acids, alternative base pairs to enlarge the genetic alphabet, or different chemccal backbones in a xenonucleic acid, these chemically modified organsims and systems will enable a much higher level of biosafety when using engineered biosystems for, or in, the environment. For example, novel enzymes (such as amylase) with non-canonical amino acids can be used to reduce the optimal temperature for breaking starch down into glucose. This would save enormous amounts of energy and help reduce greenhouse gas emissions. Synthetic biology could design organisms with an enlarged genetic alphabet or a DNA with a different chemical backbone to impede horizontal gene transfer and genetic pollution between engineered and natural organisms. Similar to protocells, xenobiology is in a very early stage of development and requires increased support for basic research in order toachieve radically new concepts and applications.

References

4.1. PROTOCELLS

Armstrong, R. (2009) Systems architecture: a new model for sustainability and the built environ using nanotech, biotech, info technology, and cognitive sci with living techno. *MIT Artif. Life*, **16**, 1–15.

Armstrong, R., and Spiller, N. (2009) Systems architecture: living buildings' special edition. *Technoetic Arts*, **7** (2), 40–43.

Barringer, F. (2009) White roofs catch on as energy cost cutters. The New York Times, http://www.nytimes.com/2009/07/30/science/earth/30degrees.html (accessed 28 May 2010).

Devlin, H. (2009) Living in the city: buildings that eat carbon dioxide? Fish bacteria that light the streets? Meet the architects rebuilding our future. Times online, http://www.timesonline.co.uk/tol/news/science/eureka/article6861966.ece (accessed 28 May 2010).

Gibson, D.G., Glass, J.I., Lartigue, C., Noskov, V.N., Chuang, R.Y., Algire, M.A., Benders, G.A., Montague, M.G., Ma, L., Moodie, M.M., Merryman, C., Vashee, S., Krishnakumar, R., Assad-Garcia, N., Andrews-Pfannkoch, C., Denisova, E.A., Young, L., Qi, Z.Q., Segall-Shapiro, T.H., Calvey, C.H., Parmar, P.P., Hutchison, C.A., III, Smith, H.O., and Venter, J.C. (2010) Creation of a bacterial cell controlled by a chemically synthesized genome. *Sciente*, **329** (5987), 52–56.

New Scientist (2010) Paint away the carbon dioxide, http://www.newscientist.com/article/mg20527424.400-paint-away-the-carbon-dioxide.html (accessed 28 May 2010).

Prize Capital LLC (2010) Carbon, www.prizecapital.net/carbon/index.html (accessed 28 May 2010).

Rasmussen, S., Bailey, J., Boncella, J., Chen, L., Collis, G., Colgate, S., DeClue, M., Fellermann, H., Goranovic, G., Jiang, Y., Knutson, C., Monnard, P.A., Mouffouk, F., Nielsen, P., Sen, A., Shreve, A., Tamulis, A., Travis, B., Weronski, P., Woodruff, W., Zhang, J., Zhou, X., and Ziock, H. (2008) *Assembly of A Minimal Protocell*, MIT press. 2008b (6), pp. 125–156.

The Green Optimistic (2010) Extremophile bacteria transform CO_2 and light into fuel better than photosynthesis, http://www.greenoptimistic.com/2009/12/11/extremophile-bacteria-transform-co2-and-light-into-fuel-better-than-photosynthesis/ (accessed 28 May 2010).

4.2. XENOBIOLOGY

Anderson, J.C., Wu, N., Santoro, S.W., Lakshman, V., King, D.S., and Schultz, P.G. (2004) An expanded genetic code with a functional quadruplet codon. *Proc. Natl. Acad. Sci. U. S. A.*, **101** (20), 7566–7571.

Bedau, M.A., McCaskill, J.S., Packard, N.H., and Rasmussen, S. (2010) Living Technology: exploiting life's principles in technology. *Artif. Life*, **16**, 89–97.

Berg, P. (2008) Meetings that changed the world: Asilomar 1975: DNA modification secured. *Nature*, **455** (7211), 290–291.

Berg, P., Baltimore, D., Brenner, S., Roblin, R.O., 3rd, and Singer, M.F. (1975) Asilomar conference on recombinant DNA molecules. *Science*, **188** (4192), 991–994.

Budisa, N. (2004) Prolegomena to future efforts on genetic code engineering by expanding its amino acid repertoire. *Angew. Chem. Int.*, **43**, 3387–3428.

Carlson, R. (2003) The pace and proliferation of biological technologies. *Biosecur. Bioterror.*, **1** (3), 203–214.

Carr, P., and Church, G. (2009) Genome engineering. *Nat. Biotechnol.*, **27** (12), 1151–1162.

Chin, J.W. (2006) Modular approaches to expanding the functions of living matter. *Nat. Chem. Biol.*, **2** (6), 304–311.

Gaskell, G., Allum, N., Wagner, W., Kronberger, N., Torgersen, H., Hampel, J., and Bardes, J. (2004) GM foods and the misperception of risk perception. *Risk Anal.*, **24** (1), 185–194.

Herdewijn, P., and Marliere, P. (2009) Toward safe genetically modified organisms through the chemical diversification of nucleic acids. *Chem. Biodivers.*, **6** (6), 791–808.

Kershner, R.J., Bozano, L.D., Micheel, C.M., Hung, A.M., Fornof, A.R., Cha, J.N., Rettner, C.T., Bersani, M., Frommer, J., and Rothemund, P.W. (2009) Placement and orientation of individual DNA shapes on lithographically patterned surfaces. *Nat. Nanotechnol.*, **4** (9), 557–561.

Leconte, A.M., Hwang, G.T., Matsuda, S., Capek, P., Hari, Y., and Romesberg, F.E. (2008) Discovery, characterization, and optimization of an unnatural base pair for expansion of the genetic alphabet. *J. Am. Chem. Soc.*, **130** (7), 2336–2343.

Luisi, P.L., Chiarabelli, C., and Stano, P. (2006) From never born proteins to minimal living cells: two projects in synthetic bio. *Orig. Life Evol. Biosph.*, **36** (5–6), 605–616.

Marliere, P. (2009) The farther, the safer: a manifesto for securely navigating synthetic species away from the old living world. *Syst. Synth. Biol.*, **3** (1–4), 77–84.

Neumann, H., Wang, K., Davis, L., Garcia-Alai, M., and Chin, J.W. (2010) Encoding multiple unnatural amino acids via evolution of a quadruplet-decoding ribosome. *Nature*, **464**, 441–444.

Ran, T., Kaplan, S., and Shapiro, E. (2009) Molecular implementation of simple logic programs. *Nat. Nanotechnol.*, **4**, 6.

Schmidt, M. (2008) Diffusion of synthetic biology: a challenge to biosafety. *Syst. Synth. Biol.*, **2** (1–2), 1–6.

Schmidt, M. (2010) Xenobiology: a new form of life as the ultimate biosafety tool. *BioEssays*, **32** (4), 322–333.

Schmidt, M., Kelle, A., Ganguli-Mitra, A., and deVriend, H. (2009) *Synthetic Biology: the Technoscience and Its Societal Consequences*, Springer, New York.

Sismour, A.M., and Benner, S.A. (2005) Synthetic biology. *Nat. Rev. Genet.*, **6**, 533–543.

Yang, Z., Hutter, D., Sheng, P.P., Sismour, A.M., and Benner, S.A. (2006) Artificially expanded genetic information system: a new base pair with an alternative hydrogen bonding pattern. *Nucleic Acids Res.*, **34** (21), 6095–6101.

Yang, Z., Sismour, A.M., Sheng, P., Puskar, N.L., and Benner, S.A. (2007) Enzymatic incorporation of a third nucleobase pair. *Nucleic Acids Res.*, **35** (13), 4238–4249.

Further Reading

Bedau, M.A., Parke, E.C., Tangen, U., and Hantsche-Tangen, B. (2009) Social and ethical checkpoints for bottom-up synthetic biology or protocells. *Syst. Synth. Biol.*, **3** (1–4), 65–75.

5
Regulatory Frameworks for Synthetic Biology

Lei Pei, Shlomiya Bar-Yam, Jennifer Byers-Corbin, Rocco Casagrande, Florentine Eichler, Allen Lin, Martin Österreicher, Pernilla C. Regardh, Ralph D. Turlington, Kenneth A. Oye, Helge Torgersen, Zheng-Jun Guan, Wei Wei, and Markus Schmidt

In the final chapters of this book, we take a look at the regulatory framework that applies to synthetic biology. These chapters will provide great insight into laws, regulations, guidlines and code of conducts that apply to synthetic biology (SB) and shape the research and development activities in this field. Taking into account the regulatory framework is crucial, especially when dealing with the kind of applications discussed in the previous chapters, many of which involve environmental safety or even biosecurity issues at one point or the other.

Since SB is clearly a global endeavor we present three chapters on three different parts of the world, namely the United States,[1] Europe and China. In addition to an analysis of existing regulations (as of 2011) we also sum up the challenges SB will pose to the different regulatory frameworks.

5.1
United States of America

5.1.1
Introduction

The field of synthetic biology encompasses the design and construction of new biological parts, devices, and systems, as well as the re-design of existing, natural biological systems for useful purposes. Although synthetic biology is a relatively

1) In 2008, Jennifer Byers-Corbin and Rocco Casagrande wrote a concise guide to US regulations for a SynBERC Biosafety, Security and Preparedness Workshop. In 2009, Kenneth Oye revised the guide for posting on the SynBERC website and use in iGEM team advisor training. In 2011, Shlomiya Bar-Yam, Florentine Eichler, Martin Oesterreicher, Kenneth Oye, Pernilla Regardh and Ralph Donald Turlington updated the US review, added reviews of the Cartagena Protocol, UN Bioweapons Convention and Australia Group guidelines and added material on EU Directives for EU–US comparisons.

Synthetic Biology: Industrial and Environmental Applications, First Edition. Edited by Markus Schmidt.

new field, regulations applied to more traditional biological research already apply to synthetic biology and its products. This regulatory landscape was brought about by the advent of recombinant DNA technology, which has been used in research for decades. Most, but clearly not all, synthetic biology research involves the use of recombinant DNA or completely synthetic DNA that is identical to DNA from existing organisms. This review treats United States regulations governing synthetic biology, the real world implementation of these regulations, and several treaties and agreements to which the United States is a party. The conclusions compare United States practices with those of the European Union and discuss future considerations for the regulation of synthetic biology.

5.1.2
United States Federal Regulations and Guidelines

This section summarizes United States Federal regulations and guidelines that affect those working in the area of synthetic biology and discuss their applicability and penalties. Although these regulations exist on paper, not all are enforced and some are not enforceable. Furthermore, the NIH guidelines, which are the most comprehensive set of guidelines for experiments involving synthetic biology, are not binding on all individuals working in the field of synthetic biology. Accordingly, after summarizing each of the regulations and guidelines, we discuss the level of enforcement and the repercussions associated with non-compliance.

5.1.2.1 National Institutes of Health: Guidelines for Research Involving Recombinant DNA Molecules[2]

While they are not always invested with binding regulatory authority, the National Institutes of Health Guidelines for Research Involving Recombinant and Synthetic Nucleic Molecules (NIH Guidelines) are perhaps the most relevant feature of the regulatory landscape, as they are the regulations with which the vast majority of researchers are most familiar. The NIH Guidelines for working with recombinant DNA[3] are intended to specify safety practices and containment procedures including the creation and use of organisms and viruses containing recombinant DNA defined as: "(i) molecules that are constructed outside living cells by joining natural or synthetic DNA segments to DNA molecules that can replicate in a living cell, or (ii) molecules that result from the replication of those described in (i) above."[4] Institutions involved in conducting or sponsoring any recombinant DNA

2) Information in this section was taken from "NIH Guidelines for Research Involving Recombinant DNA Molecules" April 2002 and the May 2011 revisions unless otherwise noted. See http://oba.od.nih.gov/oba/rac/Guidelines/NIH_Guidelines.htm accessed November 1, 2011.

3) Defined as either "molecules that are constructed outside living cells by joining natural or synthetic DNA molecules that can replicate in a living cell, or . . . molecules that result from the replication of those described above." "NIH Guidelines for Research Involving Recombinant DNA Molecules", April 2002, Section I-B. "Definition of Recombinant DNA Molecules".

4) NIH Guidelines for research involving recombinant DNA molecules (NIH Guidelines). April 2002. http://www4.od.nih.gov/oba/rac/guidelines_02/NIH_Guidelines_Apr_02.htm#_Toc7261549

research funded in part, or whole, by the NIH are required to adhere to the guidelines. In addition, funding from other federal agencies or private sources may often be contingent on compliance with the guidelines. Indeed, the NIH Guidelines are widely regarded as *de facto* standards within the research community and are often implemented by researchers who would otherwise not be obligated to do so.

Under NIH guidelines, experiments involving recombinant DNA are classified into six categories, based upon the number of regulatory hurdles they are required to clear to receive approval. The most dangerous experiments require Institutional Biosafety Committee (IBC) approval, Recombinant DNA Advisory Committee (RAC) review *and* NIH director approval prior to initiation of a proposed experiment while non-exempt experiments are considered the least dangerous and only require IBC notice which can be given *simultaneous* to initiation of the proposed experiment. See Table 5.1 for more detailed information on experiments that fall into each of the six categories listed below.

1) Experiments that require IBC approval, RAC Review, and NIH Director approval *prior to* initiation of the proposed experiment;
2) Experiments that require NIH/OBA and IBC approval *prior* to initiation of the proposed experiment;
3) Experiments that require IBC and IRB approval and RAC review *prior to* research participant enrollment;
4) Experiments that require IBC approval *prior* to initiation of the proposed experiment;
5) Experiments that require IBC notice *simultaneous* with initiation of the proposed experiment;
6) Exempt experiments.

IBC Approval: All non-exempt experiments involving recombinant DNA require Institutional Biosafety Committee (IBC) review where they are evaluated to ensure the containment levels, facilities, procedures, practices and the training and expertise of personnel involved in the research are in compliance with the NIH guidelines. When no senior regulatory committee is involved, the IBC is also responsible for setting containment levels (as specified by the NIH) when whole plants or animals are used, reviewing recombinant DNA research to insure compliance with the NIH Guidelines and adopting emergency plans covering accidental releases and personnel contamination resulting from recombinant DNA research. It is the responsibility of the IBC to report any significant problems with, or violations of the NIH Guidelines to the appropriate institutional official.

RAC Review: The Recombinant DNA Advisory Committee (RAC) functions to provide recommendations and advice to the Director of the NIH on the conduct and oversight of potentially dangerous research involving recombinant DNA. Experiments requiring RAC review must be performed using the containment levels assigned by the committee. Furthermore RAC has the authority to approve or deny experiments considered as Major Actions (see Table 5.1) under the NIH Guidelines.

Table 5.1 Required actions under NIH guidelines, by experiment type.

Experiment type	NIH guideline section	Required actions
Experiments involving the deliberate transfer of a drug resistance trait (that could compromise the use of the drug to control disease agents in humans, animals, or plants) to microorganisms that are not known to acquire the trait naturally	Section III-A-1-a – "Major Actions"	IBC approval, RAC Review, and NIH Director approval *prior* to initiation of the proposed experiment
Experiments involving the cloning of toxin molecules with $LD_{50} < 100\,ng/kg$ body weight	Section III-B-1	NIH/OBA and IBC approval *prior* to initiation of the proposed experiment
Experiments involving the deliberate transfer of recombinant DNA, or DNA or RNA derived from recombinant DNA into one or more human research participants	Section III-C-1	IBC and IRB approval and RAC review *prior* to research participant enrollment
Experiments using Risk Groups[a] 2, 3, 4 or restricted (select) agents as host–vector systems	Section III-D-1	IBC approval *prior* to initiation of the proposed experiment
Experiments in which DNA from Risk Groups 2, 3, 4 or restricted (select) agents is cloned into nonpathogenic microbes	Section III D-2	IBC approval *prior* to initiation of the proposed experiment
Experiments involving the use of infectious DNA or RNA viruses or defective DNA or RNA viruses in the presence of helper virus in tissue culture systems	Section III-D-3	IBC approval *prior* to initiation of the proposed experiment
Experiments involving whole animals	Section III-D-4	IBC approval *prior* to initiation of the proposed experiment
Experiments involving whole plants to propagate such plants, or to use plants together with microbes or insects containing DNA or for other experimental purposes (e.g., response to stress)	Section III-D-5	IBC approval *prior* to initiation of the proposed experiment
Experiments involving more than 10 l of culture	Section III-D-6	IBC approval *prior* to initiation of the proposed experiment
Experiments involving the formation of recombinant DNA molecules containing no more than two-thirds of the genome of any eukaryotic virus	Section III-E-1	IBC notice *simultaneous* with initiation of the proposed experiment
Experiments involving whole plants not previously covered in Section III-D-5	Section III-E-2	IBC notice *simultaneous* with initiation of the proposed experiment
Experiments involving transgenic rodents	Section III-E-3	IBC notice *simultaneous* with initiation of the proposed experiment
Exempt experiments	Section III-F-1, Section III-F-2, Section III-F-3, Section III-F-4, Section III-F-5, Section III-F-6	None

a) An explanation of the NIH Risk Groups is provided below.

Table 5.2 NIH risk groups.

Risk group	Description	Examples
RG1	Agents not associated with disease in healthy adult humans	Asporogenic *Bacillus subtilis* or *B. licheniformis*
RG2	Agents associated with human disease which is rarely serious and for which preventive or therapeutic interventions are *often* available	*Listeria, Staphylococcus aureus, Microsporum, Entamoeba histolytica,* mumps virus
RG3	Agents associated with serious or lethal human disease for which preventive or therapeutic interventions *may be* available	*Yersinia pestis, Histoplasma capsulatum,* yellow fever virus, TME agents
RG4	Agents likely to cause serious or lethal human disease for which preventive or therapeutic interventions are *not usually* available	Ebola virus, Lassa virus, equine morbillivirus

The NIH has set up risk groups for microbial agents used in recombinant DNA research. These risk groups classify agents according to their relative pathogenicity for healthy human adults. An explanation of each risk group is given in Table 5.2. Using these risk group classifications the NIH has put forth guidelines specifying appropriate safety precautions that must be taken when using recombinant DNA with each group of agents. These guidelines are discussed briefly below.[5)]

The guidelines regarding the NIH Risk Groups require that experiments involving the introduction of recombinant DNA into organism belonging to Risk Group 2, Risk Group 3, or Risk Group 4 (including defective viruses in the presence of a helper virus) be conducted at biosafety level (BSL) 2, level 3, and level 4 respectively. Experiments involving the transfer of DNA from Risk Group 2 or 3 agents into a non-pathogenic (Risk Group 1) organism may be performed under BSL2 containment. Importantly, the NIH guidelines state that, when transferring DNA from Risk Group 4 agents into nonpathogenic organisms, experiments should be performed under BSL4 containment until demonstration that "only a totally and irreversibly defective fraction of the agent's genome is present in a given recombinant" at which time containment can be downgraded to BSL2. Furthermore, the NIH recommends that experiments that are likely to enhance pathogenicity or extend the host range of viral vectors under conditions that permit a productive infection should be evaluated further and consideration given to increasing the BSL level by at least one.

Although the NIH risk groups encompass only microbial agents, guidelines also exist for experiments involving whole plants and animals. When experiments

5) A full list of NIH guidelines as well as exemptions can be found at: http://www4.od.nih.gov/oba/RAC/guidelines_02/NIH_Guidelines_Apr_02.htm

involve whole animals in which the animal's genome has been altered or which involve viable microbes containing recombinant DNA (other than viruses that are only vertically transmitted) that are tested on whole animals, a minimum containment of BSL2 or BSL2-N[6] is required. However, cases where the introduction of recombinant DNA into animals might lead to the creation of novel mechanisms or increased transmission of a recombinant pathogen or production of undesirable traits in a host animal containment conditions should be tightened. When experimenting with animals that contain sequences from viral vectors that do not lead to transmissible infection as a result of complementation or recombination in the host animal, BSL1 or BSL1-N containment may be used.

The NIH recommends BSL3-P or BSL2-P+[7] biological containment for many experiments involving recombinant DNA and whole plants. Experiments under this containment level include: those involving "most exotic infectious agents with recognized potential for serious detrimental impact on managed or natural ecosystems when recombinant DNA techniques are used, infectious agents with recognized potential for serious detrimental effects on managed or natural ecosystems" in which there exists "the possibility of reconstituting the complete and functional genome of the infectious agent by genomic complementation *in planta*" and, experiments with "microbial pathogens of insects or small animals associated with plants if the recombinant DNA-modified organism has a recognized potential for serious detrimental impact on managed or natural ecosystems."[8] BSL3-P containment is recommended for experiments involving sequences encoding "potent vertebrate toxins introduced into plants or associated organisms" while BSL4-P containment should be used when working with readily transmissible exotic infectious agents that have "the potential of being serious pathogens of major United States crops" when in the presence of their specific vectors.[9]

Notably, under the NIH Guidelines, synthetic DNA segments which are likely to be expressed to yield a potentially harmful polynucleotide or polypeptide (e.g., a toxin or a pharmacologically active agent) are considered as equivalent to their natural DNA counterpart. However, if the synthetic DNA segment is not expressed *in vivo* as a biologically active polynucleotide or polypeptide product, it is exempt from the NIH Guidelines. Other NIH exempt experiments include: (i) "those that

6) BSL2 and BSL2-N refer to biosafety level 2 and biosafety level 2–animals respectively. Standard practices for BSL2 can be found at: http://www4.od.nih.gov/oba/RAC/guidelines_02/Appendix_G.htm and standard practices for BSL2-N can be found at http://www4.od.nih.gov/oba/RAC/guidelines_02/Appendix_Q.htm#_Toc7256324

7) BSL2-P and BSL3-P refer to biosafety level 2 plants and biosafety level 3 plants respectively. Standard practices for BSL2-P can be found at: http://www4.od.nih.gov/oba/RAC/guidelines_02/Appendix_P.htm#_Toc7255954 and standard practices for BSL3-P can be found at: http://www4.od.nih.gov/oba/RAC/guidelines_02/Appendix_P.htm#_Toc7255970

8) Section III-D-5-e of the April 2002 revisions of the *NIH Guidelines for Research involving Recomvinant DNA molecules*. http://www4.od.nih.gov/oba/RAC/guidelines_02/NIH_Guidelines_Apr_02.htm

9) Section III-D-5-c of the April 2002 revisions of the *NIH Guidelines for Research involving Recomvinant DNA molecules*. http://www4.od.nih.gov/oba/RAC/guidelines_02/NIH_Guidelines_Apr_02.htm

consist entirely of DNA segments from a single non-chromosomal or viral DNA source, though one or more of the segments may be a synthetic equivalent", (ii) "those that consist entirely of DNA from a prokaryotic host including its indigenous plasmids or viruses when propagated only in that host (or a closely related strain of the same species), or when transferred to another host by well established physiological means", (iii) "those that consist entirely of DNA segments from different species that exchange DNA by known physiological processes, through one or more of the segments may be a synthetic equivalent" and (iv) "those that do not present a significant risk to health or the environment, as determined by the NIH Director, with the advice of the RAC, and following appropriate notice and opportunity for public comment."[10] Under the guidelines, exact copies of dangerous genes are not covered if not made by recombinant methods. This gap is addressed through proposals for reform treated at the end of this section.

In March 2009 and April 2010, NIH published proposals to update the NIH Guidelines to cover explicitly research using synthetic DNA. These actions were a result of the NSABB recommendation that the United States Government examine the language and implementation of current biosafety guidelines to ensure that they cover research with chemically synthesized DNA. The proposed amendments would broaden the scope of the Guidelines, which currently cover DNA molecules that are created via recombinant techniques, to encompass nucleic acids that are synthesized without the use of recombinant techniques. The name of the Guidelines would also be changed to "NIH Guidelines for Research Involving Recombinant and Synthetic Nucleic Acid Molecules."[11] The revised proposal published in April 2010 would substantially modify Section III-E-1. The proposal would "allow containment to be lowered provided the ability of the virus to replicate has been irreversibly impaired by a complete deletion in one or more capsid, envelope or polymerase genes required for viral replication. A quantitative criterion based on the amount of the genome present would also be maintained." The proposed revision recognizes the importance of intrinsic strategies of containment and creates incentives for the redesign of organisms to limit replication.[12]

Applicability and Enforcement of NIH Guidelines: While difficult to enforce, many of the NIH guidelines regarding the use of recombinant DNA molecules are followed by investigators wishing to comply. Although some gray areas surrounding the guidelines exist, the NIH director may aid in the interpretation of NIH guidelines for experiments not specifically addressed by the guidelines. Note that although most experiments involving synthetic biology involve the manipulation of DNA and therefore may be subject to NIH regulation, many experiments involving synthetic biology would be classified as exempt. Noncompliance with NIH guidelines may result in penalties for both the violator and the institution that supports

10) A list of natural exchangers that are exempt from the NIH guidelines can be found at: http://www4.od.nih.gov/oba/RAC/guidelines_02/APPENDIX_A.htm

11) For current NIH guidelines and proposed revisions, see http://oba.od.nih.gov/rdna/nih_guidelines_oba.html, accessed on November 1, 2011.

12) http://oba.od.nih.gov/oba/rac/News/Section%20IIIE1_NewProposal.pdf, accessed November 10, 2011.

the violator even if the violator is not a recipient of NIH funding. Penalties may include suspension, limitation, or termination of financial assistance for the violators NIH-funded research projects and similar penalties for other NIH-funded recombinant DNA research at the same institution. Alternatively researchers at the violator's institution may be required to obtain prior NIH approval of any or all recombinant DNA projects at the institution. As mentioned above, these guidelines are not enforceable for researchers that are not affiliated with an institution that receives NIH funding, furthermore much of the enforcement of these guidelines are left up to the individual institution's biosafety committee.

5.1.2.2 **Environmental Protection Agency, US Department of Agriculture and Food and Drug Administration**

NIH guidelines address research involving recombinant DNA. By contrast, EPA, USDA and FDA regulate the use and commercial production of genetically modified microbes, plants, food and drugs. Unlike the NIH guidelines, the EPA, USDA and FDA regulations apply primarily to the products of synthetic biology research that will be used for commercial purposes.

Environmental Protection Agency[13] The EPA regulates the development and production of "new"[14] microbes "for commercial purposes" created via recombinant genetics under authority from the Toxic Substances Control Act (TSCA) (15 U.S.C. 2615).[15] Anyone intending to "manufacture, import, or process" microorganisms for commercial purposes is required to file either a Microbial Commercial Activity Notice (MCAN) or a TSCA Experimental Release Application (TERA) which are used when a specific test involving release of the microorganism into the environment is planned. MCAN's must be filed with the EPA at least 90 days before use of the microbe at which time the EPA has 90 days to review the submission in order to determine whether the new microorganism may present an unreasonable risk to human health or the environment. Research that is conducted without a connection to commercial funding and is in compliance with the NIH guidelines does not require a MCAN be filed but does require a TERA application if the researcher plans to conduct a field test.[16] TERA applications must be filed with the EPA at least 60 days prior to initiating field trials.[17] "Commercial Purposes" are defined to include any activities "with the purpose of obtaining an immediate

13) Unless otherwise noted information in this section came from: Environmental Protection Agency 40 CFR Parts 700, 720, 721, 723, and 725 Microbial Products of Biotechnology; Final Regulation under the Toxic Substances Control Act; Final Rule. April, 11, 1997. http://www.epa.gov/fedrgstr/EPA-TOX/1997/April/Day-11/t8669.pdf

14) Defined as "those microorganisms formed by deliberate combinations of genetic material from organisms classified in different taxonomic genera" at the United States Regulatory Agencies Unified Biotechnology Website: http://usbiotechreg.nbii.gov/roles.asp

15) Interview with Thomas Crosetto of EPA TSCA Region 5 (IL,IN,MI,MN,OH,WI), July 26, 2007.

16) Personal communication with Jim Alwood, Chemical Control Division, EPA. February 19. 2008.

17) 40 CFR Section 725.1.

or eventual commercial advantage for the manufacturer, importer, or processor" and covers the usage of "any amount of a microorganism or microbial mixture". Commercial distribution, including test marketing, product research, and development of an intermediate are also covered by these regulations.

Notably, the regulations also apply to "substances that are produced coincidentally during the manufacture, processing, use, or disposal of another microorganism or microbial mixture, including byproducts that are separated from [it] . . . and impurities that remain in [it]." Furthermore "mobile genetic elements", defined as any "element of genetic material that has the ability to move genetic material within and between organisms. . . . Includ[ing] all plasmids, viruses, transposons, insertion sequences, and other classes of elements with these general properties" are regulated as well.[18)]

Entities making MCAN or TERA submissions to the EPA under these regulations are required to submit information allowing the microorganism to be "accurately and unambiguously identified," including taxonomic designations "for the donor organism and the recipient microorganism to the level of strain, as appropriate. These designations must be substantiated by a letter from a culture collection, literature references, or the results of tests conducted for the purpose of taxonomic classification." The submitting entity is furthermore required to provide, upon the EPA's request, data supporting the taxonomic designation, including "the genetic history of the recipient microorganism . . . documented back to the isolate from which it was derived."[19)] Submitters are moreover directed to provide "supplemental" information incorporating both phenotypic[20)] and genotypic[21)] information.[22)]

As with the NIH guidelines, the EPA regulations have also introduced certain exemptions. These exemptions apply to experiments regulated by another federal agency and those contained within a structure such as a greenhouse[23)] if researchers maintain records demonstrating eligibility. Most academic researchers are exempt from this record keeping requirement provided their institution is in compliance with NIH Guidelines for research involving recombinant DNA molecules. Likewise, the EPA has determined some organisms to be associated with low risk with

18) 40 CFR Section 725.3.
19) 40 CFR Section 725.12 (a).
20) Phenotypic information is defined as "pertinent traits that result from the interaction of a microorganism's genotype and the environment in which it is intended to be used and may include intentionally added biochemical and physiological traits". 40 CFR Section 725.12 (b) (1).
21) Genotypic information is defined as "the pertinent and distinguishing genotypic characteristics of a microorganism, such as the identity of the introduced genetic material and the methods used to construct the reported microorganism. This also may include information on the vector construct, the cellular location, and the number of copies of the introduced genetic material." 40 CFR Section 725.12 (b) (2).
22) 40 CFR Section 725.12 (b).
23) In personal communications with Jim Alwood, Chemical Control Division, EPA. February 19, 2008, it was stated that a greenhouse is not considered to be acceptable for containment of microbes.

respect to the characteristics of the recipients. These organisms are eligible for Tier I or Tier II exemptions given the users certify that they meet certain eligibility requirements. To qualify for an exemption, research must be conducted using one of the ten recipient microorganisms listed as exempt and the genetic material must meet certain criteria for toxicity, stability and poor mobilization. Furthermore, the researcher must certify they meet containment procedures, including killing the microorganism. If exempt there is no review by the EPA; the manufacturer need only certify that they meet these requirements.[24)] This exemption is most commonly used by manufactures of specialty and commodity chemicals, particularly industrial enzymes.[25)] "These exemptions could be very relevant to synthetic biology. Any manufacturer who inserts synthetic DNA into one of the eligible microorganisms and meets other criteria could use this exemption instead of an MCAN."[26)]

In addition to commercial microbes, the EPA also regulates pesticides, including genetically engineered pesticides under the Federal Insecticide, Fungicide, and Rodenticide Act (FIFRA). EPA notes "With regard to biotechnology, EPA's jurisdiction under FIFRA covers regulation of the new substance and DNA in the plant when it is pesticidal in nature"[27)] such as the introduction of gene that codes for the toxin from *Bacillius thuringiensis* into maize. Researchers wishing to gather data necessary to grant registration under Section 3 of FIFRA for a pesticide not registered with the Agency or a new use of a registered pesticide may apply for an Experimental Use Permit. Within 120 days of receiving the application and all supporting data the EPA must grant or deny the application.[28)] Before a pesticide can be marketed in the United States, the EPA must evaluate the product for possible risks to human health, risks to nontarget organisms and environment, potential for gene flow and the need for insect resistance management plans.

Applicability and Enforcement of EPA Regulations: The EPA conducts both regularly scheduled and surprise inspections of MCAN- and TERA-filing companies. According to interviews with EPA officials, the agency currently employs approximately 30 inspectors each in four of the EPA TSCA Biotechnology program's 10 regional offices (2, 4, 5, 8).[29),30)] Every MCAN or TERA-filing company has been inspected at least once. Inspections however, are restricted by statute (15 U.S.C. Section 2610) from extending to "financial data, sales data (other than shipment data), pricing data, personnel data, or research data" (other than data required

24) Personal communication with Jim Alwood, Chemical Control Division, EPA. February 22, 2008.
25) FACT SHEET Microbial Products of Biotechnology: Final Regulations under the Toxic Substances Control Act. http://www.epa.gov/oppt/biotech/pubs/pdf/fs-002.pdf
26) Personal communication with Jim Alwood, Chemical Control Division, EPA. February 22, 2008.
27) US Environmental Protection Agency: EPA's Regulations of Biotechnology for Use in Pest Management. http://www.epa.gov/pesticides/biopesticides/reg_of_biotech/eparegofbiotech.htm
28) Federal Insecticide, Fungicide, and Rodenticide Act (As Amended Through LL. 110-94, enacted October 9, 200).
29) Interview with Michael Bias, CBI Coordinator, EPA TSCA Region 2 (NJ, NY, PR, VI) on July 19, 2007.
30) EPA TSCA Biotechnology Program Contacts (very out of date) are available here: http://www.epa.gov/opptintr/biotech/pubs/biocontx.htm

to verify compliance). In the event that violations are discovered in the course of an inspection, companies may be subject to civil penalties of up to US$ 25 000 for each violation of the regulations.[31] Note that there is no assurance that nonregistered facilities are in compliance as these facilities are not subject to inspection. Although this regulation covers many research activities performed using private funding, if research is conducted without a connection to commercial funding and is in compliance with the NIH guidelines, the EPA only needs to be involved if the researcher plans to conduct a field test.[32]

5.1.2.3 USDA Animal and Plant Heath Inspection Service[33]

The Animal and Plant Health Inspection Service (APHIS) of the USDA regulates the introduction (importation, interstate movement, and release into the environment) of genetically engineered organisms (including plants, insects, microbes) that may pose a risk to plant or animal health. Depending on the plant, introduction may require an APHIS permit or APHIS notification.

Notification is a streamlined version of the permit process but cannot be used for the introduction of all modified plants, nor can it be used for other introduction of other organisms regulated by APHIS. To make use of the notification procedure the release must be fully terminated in one year and plants being introduced must meet the eligibility criteria listed in Table 5.3. Furthermore, introduction must be in accordance with all six specified performance standards found in Table 5.4. If

Table 5.3 APHIS notification eligibility criteria.[a]

Criteria
The recipient organism is not listed as a noxious weed nor considered by APHIS to be a weed in the area of release.
The introduced genetic material is “stably integrated” in the plant genome.
The function of the introduced genetic material is known and its expression in the regulated organism does not result in plant disease.
The introduced genetic material does not produce an infectious or toxic entity or encode products intended for pharmaceutical or industrial use.
The introduction does not pose significant risk of creating new plant viruses.
The plant has not been modified to contain sequences from human or animal pathogens.

a) USDA’s Biotechnology Notification Process: Biotechnology Regulatory Services, March 2005 http://168.68.129.70/lpa/pubs/fsheet_faq_notice/fs_brsbiotechnotif.html

31) 15 U.S.C. Section 2615.
32) Personal communication with Jim Alwood, Chemical Control Division, EPA. February 19, 2008.
33) Unless otherwise noted, information in this section was taken from United States Department of Agriculture Animal and Plant Health Inspection Service’s Biotechnology website: http://www.aphis.usda.gov/biotechnology/index.shtml

Table 5.4 APHIS notification performance standards.[a)]

Standards
Shipping and maintenance at destination should not lead to release of viable plant into the environment.
Caution must be taken to avoid inadvertently mixing the regulated plant with non-regulated plant materials of any species which are not part of the environmental release.
Regulated plants and plant parts must be maintained in such a way that the identity of all material is known while it is in use, and the plant parts must be contained or devitalized when no longer in use.
No viable vector agents may be associated with the regulated plant.
At the conclusion of the field trial no regulated plants or offspring may persist in the environment.
Upon termination of the field test, no viable material shall remain which is likely to volunteer[b)] in subsequent seasons, or volunteers must be managed to prevent persistence in the environment.

a) USDA's Biotechnology Notification Process: Biotechnology Regulatory Services, March 2005 http://168.68.129.70/lpa/pubs/fsheet_faq_notice/fs_brsbiotechnotif.html
b) Volunteer plants result from natural propagation, as opposed to growing after being deliberately planted.

the plant does not meet the criteria for notification, or if the engineered organism is not a plant, the applicant must follow the full permitting process. Note that some industrial research in synthetic biology involving the introduction of a plant, animal or insect would require full permitting because it does not meet the third criterion for exemption from notification: "The introduced genetic material does not produce an infectious or toxic entity or encode products intended for pharmaceutical or industrial use."[34)]

The permit process is similar to the notification process however more detailed information such as how the field tests will be performed, and how the organism will be moved may be required. APHIS reviews the information provided in the permit application to ensure that the organism will not pose harm to the surrounding environment. If there is a "high risk for a new plant variety to outcross with a weedy relative,"[35)] APHIS may not authorize a field test or may, in conjunction with granting a permit, impose additional regulations to ensure the organism is handled safely and is properly confined. When enough evidence exists that a modified organism does not pose any greater risk than an equivalent unmodified organism, APHIS may grant the organism an unregulated status. Once receiving

34) USDA's Biotechnology Notification Process: Biotechnology Regulatory Services, March 2005 http://168.68.129.70/lpa/pubs/fsheet_faq_notice/fs_brsbiotechnotif.html

35) USDA's Role in Federal Regulation of Biotechnology. Jan 10, 2006. http://www.ncsl.org/programs/agri/biotechfinal.htm

this status the modified organism may be "introduced into the United States without any further APHIS regulatory oversight."[36]

Although APHIS does not regulate the use of contained transgenic organisms, the unauthorized or accidental release of such organisms violates APHIS regulations. It is the responsibility of the researcher to ensure that unauthorized releases do not occur; therefore, APHIS encourages anyone working with transgenic organisms to abide by the NIH guidelines or other similar protocols.[37]

Finally, the Federal courts have on occasion reversed USDA approvals of GM plants. For example, on August 15, 2010, a Federal court revoked USDA approval of Monsanto's Roundup resistant GM sugar beets. The court ruled that USDA had failed to consider fully potential environmental risks of GM sugar beets, especially the risk of out-crossing through horizontal gene transfer. It is not surprising that the courts and regulatory agencies differ in weighting of risks under conditions of uncertainty and controversy.

Applicability and Enforcement of APHIS Regulations:[38] Compliance with APHIS regulations is sought via education and outreach to developers of genetically engineered organisms as well as though inspections of field sites. A percentage of Notification Fields are inspected annually. Moreover, all permitted fields are inspected at least once and pharmaceutical and industrial field tests are inspected up to seven times, including before and after a field trial. If the APHIS inspector concludes that the test field does not meet standards immediate corrective actions must be taken. While minor infractions may be corrected without disturbing the test plot, serious incidents (such as unauthorized or accidental releases) may require "destruction of research plots, quarantine of harvested crops, formal corrective action plans, or other long-term measures." Serious infractions or record of several small incidents, may lead to an investigation by the APHIS' Investigative and Enforcement Services (IES). Identification by IES of serious infractions may lead to civil penalties including fines up to US$ 500 000 and the possibility of criminal prosecution.

5.1.2.4 **Food and Drug Administration**[39]

The FDA regulates all plant-derived foods and feeds, including those altered using recombinant DNA techniques. Unlike the EPA and APHIS which have unique regulations specifically designed to address the area of synthetic biology, the FDA regulates food items that may be made via synthetic biology within the existing framework of acts including those relating to the safety of food products derived

36) United States Department of Agriculture Animal and Plant Health Inspection Service's Biotechnology website: http://www.aphis.usda.gov/biotechnology/submissions.shtml

37) USDA-APHIS Biotechnology Regulatory Services *User's Guide*: Notification. Biotechnology Regulatory Services Animal and Plant Health Inspection Service United States Department of Agriculture. Riverdale, Md. Feb. 2008.

38) Information in this section was taken from http://www.aphis.usda.gov/biotechnology/compliance_main.shtml unless otherwise noted.

39) Information in this section was taken from www.fda.gov/OHRMS/DOCKETS/98fr/011801a.htm, Draft Guidance, "Foods Developed through Biotechnology", Jan 17, 2001.

from new plants and those relating to food additives.[40] "FDA recognizes that whether there is a change in the legal status of a food resulting from a particular rDNA modification depends almost entirely on the nature of the modification, and that not every modification accomplished with rDNA techniques will alter the legal status of the food." Following this reasoning, the FDA does not require premarket approval for all foods developed using recombinant DNA technology. However development of the product often falls under the regulatory authority of APHIS or the EPA. Likewise, food developed using this technology does not, necessarily, require special labeling. If a substance added through recombinant DNA techniques does not differ from other approved additives no premarket food additive approval is required, although the product still must meet all safety regulations related to food products derived from new plants.

If the food produced through the use of recombinant DNA techniques "contain substances that are significantly different from, or are present in food at a significantly higher level than, counterpart substances historically consumed in food,"[41] the new substances may not be generally regarded as safe and may require regulation as a food additive. Furthermore the introduction of proteins that are potential allergens into foods where that particular allergen is not naturally found may require special labeling or be prohibited by the FDA.[42] Additionally the FDA is aided by the EPA when evaluating residual pesticides created through biotechnology in food and animal feed. The Federal Food, Drug, and Cosmetics Act requires EPA to set tolerances, or exemptions from tolerances, for the allowable residues of pesticides in food so as to ensure they do not pose a danger to human health.[43]

Although not required for food products created using recombinant DNA technologies, the FDA has made available a consultation process that allows developers to "actively consult with the FDA regarding their new plant varieties." It is the FDA's belief that the developers of all rDNA food commercially marketed in the US have consulted with the agency prior to marketing said food.

5.1.2.5 Department of Commerce Regulations

The Department of Commerce, Bureau of Industry and Security has licensing authority over dual-use items.[44] These regulations are based on the list developed

40) FDA's Policy on Foods Derived from New Plant Varieties. http://www.nal.usda.gov/pgdic/Probe/v2n3/fda.html
41) Premarket Notice Concerning Bioengineered Foods. January 18, 2001. http://www.cfsan.fda.gov/~lrd/fr010118.html
42) Premarket Notice Concerning Bioengineered Foods. January 18, 2001. http://www.cfsan.fda.gov/~lrd/fr010118.html
43) EPA's Regulation of Biotechnology for Use in Pest Management. June 2003. http://www.epa.gov/pesticides/biopesticides/reg_of_biotech/eparegofbiotech.htm
44) Dual-use biologicals controlled by Commerce under the Export Administration Act are listed under Export Control Classification Numbers (ECCNs) 1C351 ("Human and zoonotic pathogens and toxins"), 1C352 ("Animal pathogens"), 1C353 ("Genetic elements and genetically-modified organisms"), 1C354 ("Plant pathogens") and 1C360 ("Select agents not controlled under ECCN 1C351, 1C352, or 1C354) found in Supplement No. 1 to Part 774 of the Export Administration Regulations (EAR).

by the Australia Group described in Part Four of this document[45] and are enforced under the same classification system in each member nation.

Of particular interest is the control of the export of genetic elements that "contain nucleic acid sequences associated with pathogenicity" as well as those that contain nucleic acids coding for any of the toxins or any sub-units of a toxin.[46] A strict interpretation of this law would require a license for the export of any oligonucleotides or synthetic genes that contain sequences associated with pathogenicity of organisms on the control list.

It is worth noting that unlike many other portions of the EAR, the ECCNs governing the export of dangerous human, animal, and plant pathogens and toxins (along with the genetic material associated with them) remain in force regardless of the intended destination of the export. In other words, exports to Canada, the European Union, or Japan require that the exporter undergoes the same licensing as if it intended to export to China or Saudi Arabia. Furthermore, commerce regulations covering biologicals (Category 1C) have no "low value shipment" authorized.[47] That is, a shipment of pathogens or genetic material will fall under the regulations regardless of its value. For other controlled items, like manufacturing equipment, the item must be worth at least a certain value for Commerce Department Regulations to apply.

Furthermore, even if all exports of dual-use biologicals (such as genetic material "associated with pathogenicity" from ECCN-listed agents) were to immediately cease, the Department of Commerce would likely continue to exercise regulatory authority over many if not all synthetic biology research through its control over "deemed exports" of technology (broadly defined in this context to mean "specific information necessary for the 'development,' 'production,' or 'use' of a product" controlled under the EAR)[48] via the transfer of expertise to a non-resident foreign national (including those in the US under H-1B work visas). As in the case of "physical" exports, the "exporter" is required to seek an export license. "Exporters" are exempted from this requirement if the technology transfer occurs in the course of "fundamental research" that is pursued without a specific practical aim or with the intention of publication in the scientific or academic literature.[49] However, research and development conducted by private corporations, or funded by corporations, in which the findings are reviewed with the intent of controlling the results to be released in the open literature are considered proprietary and are subject to

45) "The Australia Group (AG) is an informal forum of countries which, through the harmonisation of export controls, seeks to ensure that exports do not contribute to the development of chemical or biological weapons." the Australia Group. http://www.australiagroup.net/en/index.html

46) These regulations are controlled by 1C351 a. to d., 1C352, 1C354, and 1C360, ECCN 1C353, Supplement No. 1 to Part 774 of the Export Administration Regulations.

47) Joe Chuchla, former Director, Nuclear Technology Division, Bureau of Export Administration (now Bureau of Industry and Security), Department of Commerce, personal communication, September 19, 2007.

48) http://www.bis.doc.gov/DeemedExports/DeemedExportsFAQs.html#23

49) EAR Part 772, page 4.

the licensing requirement.[50] Of note, the law governing dual-use exports under Commerce control has expired in August 2001 and the regulations are being governed by an Emergency Powers Act.

Applicability and Enforcement of Commerce Department Regulations: The issuance of an export license from the Commerce Department requires the exporter to submit all relevant information on the item to be exported and the end user and the government to complete a thorough government review of proposed export in a timely manner. The review process is to be completed within 30 days of the agencies' receipt of the export application for review. In practice, the licensing process typically takes from six to eight weeks.[51] Unless a license has been approved, the shipment of an item requiring such a license is illegal. The text of this rule implies that those involved in synthetic biology based businesses, such as companies that make and sell oligonucleotides, probably violate this rule when they ship synthetic DNA from pathogens. Customs and Border Protection has primary responsibility for the investigation of export violations and will likely be the first to investigate when an item is found to lack the necessary paperwork. In addition the Office of Export Enforcement within the Department of Commerce, Bureau of Industry and Security may also investigate and file charges for an illegal export. In either case, criminal and or civil charges may be filed and could involve large fines, prison time and/or the denial of all export privileges.

Due to the number of export applications received by BIS (about 12 000 yearly), NPC has restricted their review to a few areas of concern such as exports to the People's Republic of China. It is consequently difficult to imagine this review program being substantially extended to cover exports of dual-use biologicals such as synthetic nucleotides, total orders for which number in the tens of millions each year, even if the industry first determines which few percent of products exported match sequences from controlled pathogens.

Furthermore, informal discussions with some industrial and academic biologists, revealed that few, if any, have obtained an export license for the export of these dual use synthetic products. In fact, most scientists we interviewed had never heard of these regulations. Even if the Commerce Department were to enforce these regulations on biologicals, given the difficulty, without destructive testing, of identifying the dual-use nature of the products, it would be nearly impossible for customs inspectors to identify an illicit package unless the shipper declares the contents accurately.

5.1.2.6 Select Agent Rules

Although not specifically aimed at regulating synthetic biology, the Select Agent Rules (SAR; 42 C.F.R. Part 73, 7 C.F.R. Part 331, 9 C.F.R. Part 121[52]) contain

50) http://www.bis.doc.gov/DeemedExports/DeemedExportsFAQs.html#23

51) From conversations and correspondence with Joe Chuchla, former Director, Nuclear Technology Division, Bureau of Export Administration (now Bureau of Industry and Security), Department of Commerce, May 25, 2007–June 11, 2007.

52) http://www.selectagents.gov/resources/42_cfr_73_final_rule.pdf

language that makes them applicable to the field. These regulations, published jointly by the United States Departments of Health and Human Services (HHS) and Agriculture (USDA) in the Federal Register,[53)] draw upon authority established by Congress through the Antiterrorism and Effective Death Penalty Act of 1996,[54)] the Uniting and Strengthening America by Providing Appropriate Tools Required to Intercept and Obstruct Terrorism (USA PATRIOT) Act[55)] of 2001, and the Public Health Security and Bioterrorism Preparedness and Response Act of 2002.[56)] The primary goal of the Select Agent program is to regulate the possession, use, and transfer of a specified list of select agents and toxins[57)] that are considered potentially severe risks to human, animal or plant health, or to animal or plant products. However, the rules also apply to "genetic elements, recombinant nucleic acids, and recombinant organisms" derived from them. This formulation is subsequently defined as: "nucleic acids that can produce infectious forms of any of the select agent viruses listed in paragraph (b) of (Section 73.3)," "Recombinant nucleic acids that encode for the functional form(s) of any of the toxins listed . . . if the nucleic acids: can be expressed *in vivo* or *in vitro*, or are in a vector or recombinant host genome and can be expressed *in vivo* or *in vitro*" and "select agents and toxins . . . that have been genetically modified."[58)] Consequently, it would appear that the SAR are not intended to regulate the use, possession or transfer of synthetic genetic fragments that are unable, by themselves, to produce a functional form of a listed agent or toxin (such a DNA oligos or synthetic genes.) However synthetic biology technology used to create functional infectious agent or toxins appear to fall under the SAR.

Section 73.18 of the SAR confers upon the HHS Secretary the authority to order surprise inspections of the facilities and records of any registered entities (including the ability to copy any records relating to activities covered by these regulations), to ensure compliance with the Rules. HHS is further authorized to inspect and evaluate the premises and records of any entity applying for registration, prior to the issuance of a certificate of registration. According to recent correspondence with CDC officials, the CDC alone has conducted over 630 inspections since 2003 to ensure that entities are following appropriate safety and security measures, as spelled out in the regulations. All CDC-registered entities have been inspected at least once. In 2006, CDC conducted 242 inspections to register or re-register entities. Approximately 30 inspectors (both civil servants and contractors) are currently employed by the CDC Division of Select Agents and Toxins to perform these inspections.[59)]

53) These regulations represent the "final rule", superseding the "initial final rule" first promulgated in December 2002.

54) http://thomas.loc.gov/cgi-bin/query/z?c104:S.735.ENR:

55) http://www.selectagents.gov/resources/USApatriotAct.pdf

56) http://www.selectagents.gov/resources/PL107-188.pdf

57) Information is available at www.selectagents.gov/resources/salist.pdf

58) 42 CFR Section 73.3.

59) From Correspondence with Lori Bane, Compliance Officer, CDC Division of Select Agents and Toxins, various dates.

APHIS inspection procedures are similar to those of the CDC, though APHIS inspectors are not "centrally located within [the] Select Agent Program."[60] While CDC inspectors are based within the CDC Select Agent Program and write the inspection reports as well as perform the inspections, the approximately 50 APHIS inspectors complete standard checklists and submit them to the APHIS Select Agent Program for review. The contents of these checklists are then incorporated into inspection reports generated within the APHIS Select Agent Program.

Applicability and Enforcement of Select Agent Rules: In the event that suspicion of a safety or security violation arises, whether in course of inspections, or by some other means (such as a tip), the regulatory authorities are able to apply administrative, civil, and/or criminal penalties upon violators. Upon revocation or suspension of a certificate of registration, the entity is required to "immediately stop all use of each select agent or toxin covered by the . . . order . . . safeguard and secure each select agent or toxin . . . from theft, loss, or release, and comply with all disposition instructions issued" by the relevant lead agency.[61] Regarding civil penalties, the Office of the Inspector General of the Department within the HHS (and its analog within the USDA) is delegated authority to conduct investigations and to impose civil money penalties against any individual or entity for violations of the Select Agent Rules, as authorized by the Public Health Security and Bioterrorism Preparedness and Response Act of 2002. Civil penalties may be pursued in conjunction with criminal ones.

In the event that a suspicion of criminal misconduct arises on the part of CDC/APHIS or the Inspector s General of HHS and USDA, the case will be referred to relevant officials with the FBI.[62] Criminal penalties for violations of the Select Agent regulations are defined under Title 18 Section 175b of the United States Code. Possession without registration of a select agent or toxin is punishable by up to five years' imprisonment and/or a fine. Transfer to unregistered persons (if the transferor knows, or has reasonable cause to believe recipient is unregistered) is punishable by up to five years' imprisonment and/or fine.

Furthermore, the transport or shipment (excluding "duly authorized" US Government activity) of select agents or toxins via interstate or international commerce (or the receipt of select agents or toxins through interstate or international commerce) is punishable by up to 10 years' imprisonment and/or fine.[63] Section 175 of Title 18 of the US Code prescribes criminal penalties of fines and/or imprisonment for up to 10 years, for the possession of "any biological agent, toxin, or delivery system of a type or in a quantity that, under the circumstances, is not

60) From Correspondence with Lori Bane, Compliance Officer, CDC Division of Select Agents and Toxins, August 6, 2007.

61) 42 CFR Section 73.8.

62) Should suspicion of criminal activity arise, CDC/APHIS would be expected to immediately suspend registration and to contact the FBI through a well defined and robust liaison system. However, to the best knowledge of our contacts at the FBI WMD Directorate and the CDC Select Agent Division, suspicion of criminal misconduct has yet to arise in the course of CDC or APHIS investigations of SAR violations. (From interviews and correspondence with Lori Bane of CDC and Tracy Rice of FBI).

63) 18 U.S.C. Section 175b.

reasonably justified by a prophylactic, protective, bona fide research, or other peaceful purpose".[64] Possession of such materials for "use as a weapon" is, for its part, punishable by a fine, life imprisonment or both.[65]

Although individuals and institutions registered to possess Select Agents appear to be well regulated, synthetic biology may allow non-registered persons access to these agents. Companies that produce custom DNA to order are urged under current law to determine what they are making or to screen who they are shipping the products to as the SARs do not apply to synthetic genetic fragments that are unable, by themselves to produce a functional form of a listed agent or toxin. However, possession of such synthetic fragments could allow unauthorized individuals to create prohibited microbes or toxins. Individuals doing so would be in violation of the SARs and would therefore be subject to criminal charges, if their activity was discovered.

5.1.2.7 **Screening Guidance for Providers of Synthetic Double-Stranded DNA**

On October 13, 2010, HHS published *Screening Framework Guidance for Providers of Synthetic Double-Stranded DNA* in the Federal Register (Federal Register, Vol. 75, No. 197 2010). Following the guidance is voluntary, but it nonetheless provides insights into how the government could and might aim to limit risks to security in an age of synthetic biology. Under this guidance, providers of synthetic, double-stranded DNA accept two important responsibilities. First, they agree that they should know to whom they are distributing a product. Second, they should know whether the product they are synthesizing and distributing contains a "sequence of concern".

To achieve this, providers have agreed to conduct customer and sequence screening. The purpose of the customer screening is to establish the legitimacy of customers ordering synthetic double-stranded DNA sequences by verifying the identity and affiliation of customers and identifying any "red flags" of which there is suspicion that the order could be used for inappropriate ends. Providers agree to check the customer against several lists of proscribed entities, such as the Department of Treasury Office of Foreign Asset Control list of Specifically Designated Nationals and Blocked Persons and the Department of Commerce Denied Persons List for domestic orders, and are required to follow the laws and regulations of United States trade sanctions and export controls for international orders.

The purpose of the sequence screening is to identify whether "sequences of concern" are ordered. Sequences of concerns are defined as those that code for the select agents and toxins identified by CDC and APHIS in the SAR. If the complete sequence or unique parts of the sequences are identified, providers must make sure that customers have a certificate of registration from CDC or APHIS for using select agents or toxins. For international orders, providers should also screen for items on the Commerce Control List to ensure that they are in compliance with the Export Administration Regulations (EAR).

64) 18 U.S.C. Section 175 (b).
65) 18 U.S.C. Section 175 (c).

If either the customer or sequence screening causes concern, a follow-up screening must take place to verify the legitimacy of the customer and end use of their order. What a follow-up screening is to be composed of is less specific than for the two initial screenings, but as far as possible, the identity, affiliation and legitimacy of the customer must be obtained as well as the intended use of the ordered DNA. If the follow-up screening does not solve the concerns raised, the provider should contact the United States Government, or more specifically either the FBI, the Select Agent Programs of CDC and APHIS or the Department of Commerce for assistance and further action.

The viability of this voluntary agreement hinges on the relatively low cost of compliance with its terms and on the concentrated structure of the synthesis sector, with a relatively small number of relatively advanced firms doing most advanced synthesis. If advanced synthesis capabilities diffuse from a few firms to many and from firms to individuals, the effectiveness of voluntary codes of conduct will obviously decline.

5.1.3 International Conventions and Agreements

The United States has accepted obligations to protect biological diversity, biosafety and biosecurity as a party to international conventions and agreements. This section reviews major provisions of these agreements and discusses their potential relevance to the field of synthetic biology.

5.1.3.1 The Convention on Biological Diversity

The Convention on Biological Diversity is a legally binding international treaty that was opened for signature in the Earth Summit in Rio de Janeiro in June 1992 and entered into force on December 29, 1993. The United States has signed, but not ratified the Convention. The three main goals of the Convention on Biological Diversity are stated in Article 1:

1) Conservation of biological diversity;
2) Sustainable use of biological resources and diversity;
3) Fair and equitable sharing of the benefits arising from genetic resources.

In order to achieve this, Contracting Parties are required to, among other things, develop and implement national strategies for the conservation and sustainable use of biological diversity, identify and monitor components of biological diversity important for its conservation and sustainable use, regulate or manage biological resources, establish a system of protected areas, promote the protection of ecosystems and encourage the equitable sharing of benefits arising out of utilization of traditional knowledge. Further, developed countries are encouraged to cooperate with and share technical and scientific knowledge relevant to biological diversity with developing countries and to provide financial support for the implementation of the Convention by developing countries.

The Convention also discusses the development and use of biotechnology and genetic resources. It specifically requires Contracting Parties including the United States to "[e]stablish or maintain means to regulate, manage or control the risks associated with the use and release of living modified organisms resulting from biotechnology which are likely to have adverse environmental impacts that could affect the conservation and sustainable use of biological diversity, taking into account the risks to human health."

In relation to access to genetic resources, the Convention establishes that the authority to determine access to genetic resources rests with national governments and hence that access must rest on mutually agreed terms and subject to prior informed consent. Contracting Parties including the United States are also required to take legislative, administrative or policy measures to provide for participation in biotechnological research, the fair and equitable sharing of results and benefits arising out of biotechnologies, especially with regard to developing countries who contribute with the genetic resources for such research. Parties also have to develop a protocol for the safe handling, transfer and use of living modified organisms.

5.1.3.2 **The Cartagena Protocol on Biosafety and the Nagoya–Kuala Lumpar Supplementary Protocol on Liability**

The Cartagena Protocol provides a biosafety extension to the Convention of Biological Diversity. It was finalized and adopted in Montreal on January 29, 2000. To date, 159 countries and the European Union have ratified or acceded to it (United Nations, 2010). Its objectives are stated in Article 1 as: "to contribute to ensuring an adequate level of protection in the field of the safe transfer, handling and use of living modified organisms resulting from modern biotechnology that may have adverse effects on the conservation and sustainable use of biological diversity, taking also into account risks to human health, and specifically focusing on transboundary movements." As such, it is intended to create a uniform international procedure for regulating the safe transfer of living modified organisms.

During the drafting of the Cartagena Protocol, negotiators differed over the question of whether it should address human health risks in addition to environmental risks (Street, 2007). The current protocol contains references to human health and a specific section on GM food and feed, but these parts are weaker than those that address environmental risks. This is probably caused by the fact that governments view citizens and their wellbeing as more relevant constituents than the environment and are hence more reluctant to give up sovereignty in the politically more relevant area of human and public health than in environment.

The Cartagena Protocol only applies to international transboundary movement of living modified organisms for the intentional introduction into the environment of the importing party. Food, feed and organisms designed for contained use are partially covered by the Protocol, but not as thoroughly as organisms for intentional release. Living modified organisms are defined as any living organism that possesses a novel combination of genetic material obtained either through application of *in vitro* nucleic acid techniques (rDNA and direct injection of nucleic acids) or by the fusion of cells beyond the taxonomic family.

Before exporting living modified organisms for intentional release into the environment exporters must notify the competent authority of the importing party in writing and submit specific information. The importer must perform a scientifically sound risk assessment based on the information submitted by the exporter, identify and evaluate the possible adverse effects of the modified organisms on the conservation and sustainable use of biological diversity and then make its decision based on this assessment.

The tension within the text regarding the grounds upon which parties may make decisions on the import of GMOs is interesting, especially since there were serious disagreements about the proper scope of the Protocol during the negotiations. The discussions about scope were centered on whether human health risks should be included, whether parties should be allowed to make decisions based on socio-economic factors and whether the precautionary principle should be allowed as a basis for decision making (Street, 2007).

The issue of whether socio-economic considerations should be allowed as basis for decision making is unclear in the final text: Article 26 allows for the consideration of socio-economic factors, while Article 10 requires decisions to be based on scientifically sound risk assessments. Furthermore, the phrase "in consistence with their international obligations" included in Article 26 is a reference to their responsibilities under WTO agreements and thus an attempt to limit the ability of states to justify their decisions based on Article 26. Likewise, the precautionary principle is severely limited and may only be implemented if the scientific risk assessment is indecisive.

On October 12, 2010, parties to the Cartagena Protocol on Biosafety finalized six years of negotiations on a new treaty to establish international rules and procedures for liability and redress in case of damage to biological diversity resulting from living modified organisms (United Nations, 2010). The "Nagoya–Kuala Lumpar Supplementary Protocol on Liability and Redress" will provide legally binding rules for transboundary movements of living genetically modified organisms under the Cartagena Protocol.[66]

5.1.3.3 The Biological Weapons Convention

The Convention on the Prohibition of the Development, Production and Stockpiling of Bacteriological (Biological) and Toxin Weapons and Their Destruction was opened for signature on April 10, 1972, and entered into force on March 26, 1975. As of June 2005, 171 states had signed the convention, of which 16 still needed to ratify it, while 23 states had not signed (The Biological and Toxin Weapons Convention, 2005). It supplements the League of Nations' Geneva Protocol of 1925, which only prohibits the use of chemical and biological weapons during warfare (League of Nations, 1925) and is currently signed by 173 states.

66) See GMO Compass, 2010; United Nations, 2010.

Article I of The Biological Weapons Convention prohibits signatory states from developing, producing, stockpiling or otherwise acquiring or retaining "microbial or other biological agents, or toxins whatever the origin or method of production, of types and in quantities that have no justification for prophylactic, protective or other peaceful purposes [and] weapons, equipment or means of delivery designed to use such agents or toxins for hostile purpose or in armed conflict." Subsequent articles require States to destroy or divert to peaceful purposes already existing biological agents and toxins, not to transfer or assist other state or non-State actors to manufacture or acquire biological agents and toxins, to take the necessary measures to prevent the development, production, stockpiling, acquisition or retention of agents within their own territory and to file complaints if it finds that another State violates the convention.

5.1.3.4 The Australia Group Guidelines

The Australia Group first met in 1985 and consisted of 15 countries. The meeting was prompted by the use of chemical weapons by Iraq in the Iran–Iraq war. Iraq had developed its chemical arsenal using tools and compounds legally purchased from Western nations.[67] The Australia Group formed to prevent Iraq from acquiring materials for the production of chemical weapons through otherwise legitimate trade through the harmonization of national export controls. The Australia Group today is a collection of 41 countries[68] that have harmonized export controls over materials and technologies likely to contribute to the development of chemical or biological weapons.[69] The group provides a set of guidelines but has no enforcement mechanism. All Australia Group countries are signatories of the Chemical Weapons Convention and the Biological Weapons Convention, and "strongly support efforts under those Conventions to rid the world of Chemical and Biological Weapons."[70] The group works by consensus, the agreement is non-binding, and all implementation takes place on the national level. It is important to note that the Australia Group is an informal arrangement that operates without the use of legally binding obligations, relying heavily on the fact that all members are also members of the Biological Weapons Convention.

Biological agents and dual use biological technology were added to the Australia Group guidelines in 1992. The initial control list was published that year, and it has expanded greatly since.[71] In 2008, the group set up a synthetic biology advisory

67) Rajiv Nayan, Australia Group, CBW Magazine, July 2008, volume 1, issue 4. http://www.idsa.in/cbwmagazine/AustraliaGroup_rnayan_0708

68) At the time of writing this guide, they included: Argentina, Australia, Austria, Belgium, Bulgaria, Canada, Croatia, Cyprus, Czech Republic, Denmark, Estonia, European Union, Finland, France, Germany, Greece, Hungary, Iceland, Ireland, Italy, Japan, Republic of Korea, Latvia, Lithuania, Luxembourg, Malta, Netherlands, New Zealand, Norway, Poland, Portugal, Romania, Slovakia, Slovenia, Spain, Sweden, Switzerland, Turkey, Ukraine, United Kingdom and the United States.

69) The Australia Group, home page, http://www.australiagroup.net/en/index.html

70) Ibid, An introduction, http://www.australiagroup.net/en/introduction.html

71) Rajiv Nayan, Australia Group, CBW Magazine, July 2008, volume 1, issue 4. http://www.idsa.in/cbwmagazine/AustraliaGroup_rnayan_0708

body to keep up with developments and to suggest responses to new innovations.[72] Currently, Australia Group guidelines relating to biological components cover dual-use technology, advanced software not available to an untrained user and biological components that have pathogenic properties (bacteria, viruses, toxins, etc.). The guidelines also regulate: (i) genetic elements containing nucleic acid sequences associated with the pathogenicity of any of the microorganisms on the control list, (ii) genetic elements containing nucleic acid sequences coding for any of the toxins in the list, or for their sub-units, (iii) genetically modified organisms that contain nucleic acid sequences associated with the pathogenicity of any of the microorganisms in the list and (iv) genetically modified organisms that contain nucleic acid sequences coding for any of the toxins in the list or for their sub-units.[73] The guidelines specify: "genetically modified organisms includes organisms in which the genetic material (nucleic acid sequences) has been altered in a way that does not occur naturally by mating and/or natural recombination, and encompasses those produced artificially in whole or in part. Genetic elements include inter alia chromosomes, genomes, plasmids, transposons and vectors whether genetically modified or unmodified, or chemically synthesized in whole or in part."[74] If the synthesized components do not code for part of a controlled pathogen or toxin, the guidelines do not apply. Also, if the genetic parts do not make up part of a genetic sequence that produces pathogenicity, then the components do not fall under the guidelines of the Australia Group, even if they do come from an organism on the control list. However, as new components are synthesized or isolated, their full functions within the DNA of a controlled pathogen may not be fully understood. This can lead to accidental shipments of components that have possible pathogenic properties.

As synthetic biology advances, more research will be performed on potentially dangerous organisms. Although the Australia Group has set up an advisory board to stay up to date with advances in the field, care must still be taken to ensure that potentially dangerous components are not moved across national boundaries inappropriately. A goal of synthetic biology is to create ways to more easily to modify organisms without advanced skills and equipment. This can allow untrained or even malicious actors to easily create a dangerous organism by assembling parts acquired from many sources. Due to these rapidly developing technologies, the guidelines laid down by the Australia Group are extremely relevant, and the group's role in the international control of dangerous organisms should not be understated. The advisory body set up in 2008 should communicate often to keep up to date with this rapidly changing technology.

72) Ibid.

73) Australia Group, Biological Agents for Control. http://www.australiagroup.net/en/biological_agents.html

74) Ibid.

5.1.4
Conclusions: Current Coverage and Future Considerations

5.1.4.1 Current Coverage

How do United States (US) regulations and guidelines that bear on synthetic biology compare with European Union (EU) Directives and regulations? In comparing US and EU standards that govern synthetic biology, it is important to examine both the letter of laws and their practical applicability. Levels of stringency for environment do not necessarily hold for security, practices in Western Europe do not necessarily hold for Eastern Europe, and practices within the US may vary from agency to agency while practices in the EU may vary from State to State and region to region within States.

As a first cut, relative to the US, the EU regulatory framework for approval of genetically engineered products for human consumption or environmental release addresses a broader range of biosafety risks with higher evidentiary standards. By contrast, with respect to biosecurity and synthetic biology, the postures of the US and EU reverse. US regulations of exports and domestic activities are more extensive and rigorously enforced than corresponding EU regulations. As discussed in the previous section, the EU and US have taken steps to implement international agreements bearing on biosecurity including the United Nations Bioweapons Convention and the Australia Group guidelines but significant gaps in coverage of the Bioweapons Convention and the voluntary nature and lack of specificity of Australia group guidelines are current sources of concern.

As Table 5.5 suggests, the United States approach is premised on the assumption that regulation should focus not on production process *per se* but on the properties of products as regulated under existing statutes. Consequently, synthetic biology products are currently covered by three different US agencies operating under four separate statutes. The result is a regulatory system marked by fragmentation, lack of coordination and different standards for different types of products. For example, review of genetically modified animals that will not be used directly as food is more or less unregulated under the current framework, while genetically modified crops with plant incorporated protectants are nominally reviewed by all three agencies. Differences in statutory mandates, risk assessment methodologies and agency cultures between the EPA, APHIS and USDA create a system where the stringency of the risk assessment and approval process is more dependent on the specific product category than on the risk of the product. By contrast, EU regulations are premised on the assumption that distinctive regulations are needed to govern risks associated with recombinant DNA production methods.

In practice, differences between US and EU regulations of health and safety risks of synthetic biology may be more complex and difficult to characterize than the summary above suggests. Within the EU, national implementation of EU directives varies from nation to nation, with less rigorous regulation of Directives on contained applications and deliberate release in Spain or Estonia than in Germany or Denmark. The degree of funding of enforcement bodies and

Table 5.5 Analysis of regulatory coverage of safety and environmental risks of synthetic biology.

Risk	International regulation	US regulation	EU regulation
Transfer of genes	Cartagena protocol on biosafety	EPA and APHIS	Directive 2001/18/EC
Mutations, evolution and proliferation		EPA	Directive 2001/18/EC
Effects on ecosystem and other species	Cartagena protocol on biosafety	EPA and APHIS	Directive 2001/18/EC
Effects on biodiversity	Convention on biological diversity, cartagena protocol on biosafety		Directive 2001/18/EC
Consumption risks		EPA (only for plant incorporated pesticides)	Regulation 1829/2003
Risks to laboratory workers		NIH guidelines	Directive 2009/41/EC, Directive 2000/54/EC
Accidental release of laboratory strains		NIH guidelines	Directive 2009/41/EC

inspectorates also produces differences in effective levels of implementation. This can be seen quite clearly within as well as between countries. In Belgium, the Flemish Region employed six inspectors, the Brussels–Capital Region two and the Walloon Region had no specialized inspectors at all.[75] Finally, it is important to mention that a given Directive may have different effects on Member States. In some countries the impact of the Directive is limited, for their national legislation has already met the requirements of the Directive (e.g., see the UK and the Directive on the deliberate release into the environment of GMOs). Other Member States like Estonia might implement national laws that just satisfy the minimum requirements of the Directive. By contrast, Belgium, has extended the scope of the Directive on the contained use of GMMs to include GMOs and nonGM human, animal and plant pathogens as well. Within the US, the rationale of the US government to not enact separate statutes for biotechnology has been undermined by specific regulations promulgated by responsible agencies. The US has, in effect, adopted on a piecemeal basis a regulatory system with separate review mechanisms for most biotechnological products like the European system. Finally, the degree to which current statutes, such as the TSCA, can be stretched to include and effectively handle technologies and products of synthetic biology remains unclear. Although individuals are responsible for living up to letter of the law, differences between formal requirements and informal implementation are often significant.

75) http://www.biosafety.be/CU/PDF/Summary_report_EC_2009.pdf

5.1.4.2 Future Prospects

The regulations summarized above deal with the use, sale and transport of synthetic biology products and processes. Many of these regulations are in their infancy. Where regulations exist, they often focus on the use of recombinant DNA or the transfer or use of microbes (or microbes derived from them) from set lists. That being said, synthetic biology challenges the utility of lists. Although it may seem like the height of sophistry, these laws do not define what *Bacillus anthracis* is and how it differs from *B. cereus*. Both can cause illness in people (albeit *B. anthracis* causes more serious illness) so fall under the definition of a biological agent. Potentially, a researcher can modify *B. cereus* by adding into it all the genes that *B. anthracis* has that *B. cereus* does not and deleting from it all genes that *B. cereus* has and *B. anthracis* does not. Since the genes from *B. anthracis* themselves are not regulated (not even the genes that encode anthrax toxins are listed, unlike in other toxogenic bacteria) and this modified organism is not technically *derived* from *B. anthracis*, this organism probably would not be considered *B. anthracis* even if, after all manipulations, the final strain resembled *B. anthracis* more than any other organism. This situation becomes even more complex if the researcher was careful not to modify the *B. cereus* strain with *exact* copies of *B. anthracis* genes, but used modified genes that encoded the same amino acid sequence via a different DNA sequence. Although this example strains the boundaries of scientific possibility, similar experiments with viruses on the list (and their close relatives) can be achieved in several laboratories today. Some regulations depend on the behavior of the agent partially address this problem; for example, the NIH guidelines classify agents by their ability to cause disease in humans. For completely novel agents or for highly modified versions of existing agents it is difficult to determine their place on this list, especially for agents that cause disease only in humans.

Although the above discussion primarily pertains to those trying to subvert list-based controls and not legitimate researchers, the action of government to prevent this subversion could lead to more burdensome regulations that may interfere with research in synthetic biology. For example, some commenters have suggested that viral select agents be defined by sharing a quantum of genetic information with (presumably canonical) viruses on the list. For example, one commenter on the Select Agent Rule suggested controlling any synthetic or natural piece of nucleic acid "comprising at least 15% of the genome of the select agent."[76] The government wisely chose not to adopt this suggestion because it would regulate viral fragments founds in several laboratories performing basic molecular investigations of viruses. Similar problems arose when the regulations attempted to produce a molecular definition of *variola major*.[77] Another proposal is to define an organism using a few genes that are potentially dangerous and regulating all organisms with these genes. This system too has flaws. We are not currently able to identify all genes associated with pathogenicity or how genes work together to cause harm. In many cases a gene that causes pathogenicity in "organism A" may

76) Federal Register, volume 70, issue 52, p. 13298; March 18, 2005.
77) Enserink, M. (2005) Unnoticed amendment bans synthesis of smallpox virus. *Science* **307**:5715.

be harmless when inserted into "organism B." If organisms are defined by a few genes, it is likely that many harmless bacteria will be subject to regulation. Furthermore, because a gene may only be recognized as being critical for pathogenicity after the writing of a rule, the rules must either be constantly updated or become quickly obsolete.

Further complicating the issue of regulation is the possibility of development of novel totally synthetic organisms. In May 2010, the J. Craig Venter Institute announced the stepwise creation of a synthetic bacterial chromosome based on *Mycobacterium mycoides* and its transfer into *M. capricolum*, where it replaced the native DNA. The synthetic microbial cell began replicating and making new sets of proteins. The properties of the synthetic cell are almost identical to natural *M. mycoides*. Over the long term, similar approaches will be used to produce synthetic organisms of increasingly novel design.[78] Although at the time this manuscript was written such a novel entirely synthetic organism has not been constructed, the possibility for creating such an organism exists. The regulation of organisms that do not yet exist is fraught with challenges. Should all unique synthetic organisms be regulated or only potentially dangerous organisms? How can one determine if a unique synthetic organism is dangerous? Because the introduction of any new organism into an existing ecosystem can have adverse effects on indigenous species, the introduction of a seemingly harmless synthetic organism into the environment could have similar effects.

Artificiality poses many novel regulatory challenges. For example, synthetic organisms may possess the ability to transfer genetic material, assuming they have any that can be transferred, with nonsynthetic organisms. Existing regulations governing release of GMOs in the United States and the European Union explicitly raise the potential for horizontal gene flow as a major consideration in assessment of risks of release. One approach to reducing risks of gene flow entails knocking out codons to create increasingly artificial versions of the microorganisms that serve as biological chassis for implanted pathways. The paradox? The increasing artificiality of microorganisms that is intended to limit gene flow also triggers more regulatory scrutiny. The problem of developing methods for appraising risks and benefits associated with increasing novelty has yet to be addressed.

Synthetic biology has great promise for curing diseases, creating fuels and solving countless other problems we currently face. Unfortunately, the improper or malicious use of such technology creates a need for regulation. One of the greatest challenges facing those who create such regulations will be weighing the costs and benefits of each rule and developing an effective enforcement system. Well crafted regulations should account for the changes synthetic biology will foster while not unduly hindering beneficial research. The field of synthetic biology and the assessment of associated benefits and risks characterized by rapid change, exceptional complexity and substantial uncertainty. The promotion of benefits and governance of risks will require the development of more adaptive systems of regulation with explicit recognition of uncertainty, with well defined plans to

78) http://www.sciencemag.org/content/328/5981/958.full.pdf

gather information that bears on policy relevant sources of uncertainty, and with clearly defined mechanisms for incorporating emerging information on the nature of benefits and risks into regulations, rules and guidelines. In short, designers of systems of regulation should take their lead from the qualities of the adaptive and evolving living systems that they seek to protect.[79)]

5.2 Europe

5.2.1 Introduction

Synthetic biology (SB) is a new field that promises to revolutionize biotechnology. SB, in short, is the application of engineering principles to biology. The more ambitious attempts are not only to re-engineer living systems, but also to design entirely new ones: to create life itself from nonliving materials (EASAC, 2011). Early assessments by civil society organizations (ETC) have criticized that regulation of such a powerful novel field is lacking, and that hazards might be imminent if regulation does not take up new developments in this field. Scientists in the field, however, hold against that there is a very comprehensive regulation on biotechnology that applies to synthetic biology as well and that there are few, if any, novel aspects that would not be covered by existing regulations. This view was corroborated in reports from a number of policy and research think tanks as well as bioethics councils in Europe and the United States.

In this chapter we look at the debate on regulation from several perspectives:

1) How is SB defined, why is it different (or not) form genetic engineering and which type of research and development is currently taking place?
2) Which regulatory frameworks, laws and guidelines are in place in Europe? We present a short overview over what the EU has to offer and deepen this account by looking at four examples of national regulations, two from large countries and two form smaller countries.
3) How adequate is the current set of regulations in Europe seen by various stakeholders and advisory bodies?
4) Based on the three perspectives above, what remains as important challenges to policy makers and scientists in Europe regarding the regulatory framework of SB?

79) For examples, of adaptive approaches to risk governance in other domains, see "Planned Adaptation in Risk Regulation: An Initial Survey of US Environmental, Health, and Safety Regulation," Lawrence McCray, Kenneth A. Oye and Arthur C. Petersen, *Technological Forecasting & Social Change*, volume 77, issue 6, July 2010, pp. 951–959; and "Adaptive Licensing:Taking the Next Step in the Evolution of Drug Licensing," Hans Georg Eichler, Kenneth A. Oye *et al.*, 2012, *Nature Clinical Pharmacology and Therapeutics*, under review.

5.2.1.1 Synthetic Biology as a Novel Science and Engineering Field

Although the term "synthetic biology" was already used about 100 years ago (Leduc, 1910, 1912), the contemporary version is a relatively young field at the intersection of biology, engineering, chemistry and information technology (Campos, 2009). Not untypical for an emerging science and engineering field, a variety of definitions circulate in the scientific community, and no one definition would receive total support by the researchers involved in SB activities (Schmidt and Pei, 2011). The probably least contested definition is that found at the SB community webpage[80]:

> Synthetic Biology is: (a) the design and construction of new biological parts, devices, and systems, and (b) the re-design of existing, natural biological systems for useful purposes. Synthetic biologists are currently working to:
>
> - Specify and populate a set of standard parts that have well-defined performance characteristics and can be used (and re-used) to build biological systems;
> - Develop and incorporate design methods and tools into an integrated engineering environment;
> - Reverse engineer and re-design pre-existing biological parts and devices in order to expand the set of functions that we can access and program;
> - Reverse engineer and re-design a 'simple' natural bacterium;
> - Minimize the genome of natural bacteria and build so-called protocells in the laboratory, to define the minimal requirements of living entities;
> - Construct orthogonal biological systems, such as a genetic code with an enlarged alphabet of base pairs.

The lack of a well-accepted definition does not stop the community from going ahead and do SB, naturally leading to a quite diverse area of science and engineering. Activities are currently performed in several fields. For various reasons, they are not always addressed under the term proper, while others use the label for still different endeavors (see Table 5.6). By and large, however, the following activities are usually subsumed under SB (Benner and Sismour, 2005; Luisi, Chiarabelli, and Stano, 2006; O'Malley *et al.*, 2008; Bedau *et al.*, 2009; Deplazes, 2009; Schmidt *et al.*, 2009):

- DNA synthesis (or synthetic genomics);
- Engineering DNA-based synthetic biological circuits;
- Defining the minimal genome (or minimal cell);
- Building protocells (or synthetic cells);
- Xenobiology (aka chemical synthetic biology).

80) See: syntheticbiology.org

Table 5.6 Examples of R&D done in synthetic biology according to complexity levels (rows) of the five major subfields (columns). See: (Schmidt and Pei, 2011).

	DNA synthesis	Genetic circuits[a)]	Minimal genomes	Protocells	Xenobiology
Biochemistry	–	–	–	Standard or alternative biochemistry (Rasmussen *et al.*, 2004)	Alternative biochemistry: XNA (Declercq *et al.*, 2002; Herdewijn and Marliere, 2009; Nielsen and Egholm, 1999; Schoning *et al.*, 2000), unnatural bases (Henry and Romesberg, 2003; Yang *et al.*, 2006; Yang *et al.*, 2007), amino acids (Hartman *et al.*, 2007)
Genes/parts	Synthetic genes (Carlson, 2003, 2009; May, 2009)	Genes and bioparts, bioparts (Cantor, Labno, and Endy, 2008; Endy, 2005; Smolke, 2009)	–	Engineered phospholipids (Murtas, 2010; Rasmussen *et al.*, 2004; Szostak, Bartel, and Luisi, 2001)	Changing the codon assignment of genes (Budisa, 2004; Luisi, 2007)
Biological systems	Artifical chromosomes (Carlson *et al.*, 2007; Macnab and Whitehouse, 2009), synthetic viruses (Cello, Paul, and Wimmer, 2002; Tumpey *et al.*, 2005; Wimmer *et al.*, 2009)	Enhanced metabolic engineering, bioparts and devices (Elowitz and Leibler, 2000; Lu, Khalil, and Collins, 2009; Ro *et al.*, 2006; Tigges *et al.*, 2010)	–	Cellular vesicles lacking key features of life (Hanczyc and Szostak, 2004; Mansy *et al.*, 2008; Mansy and Szostak, 2008)	Development of novel polymerase and ribosomes (Ichida *et al.*, 2005; Loakes and Holliger, 2009; Neumann *et al.*, 2010)
Organelles, single-cell organisms[b)]	Whole genome synthesis (Carr and Church, 2009; Dymond *et al.*, 2011; Gibson *et al.*, 2008; Gibson *et al.*, 2010; Lartigue *et al.*, 2007)	–	Top-down SB reducing existing organisms' genomes (Hutchison *et al.*, 1999; Mushegian, 1999; Posfai *et al.*, 2006)	Real synthetic cells, bottom-up SB manufacturing whole cells (Rasmussen, 2009)	Xeno-organisms, chemically modified organisms (CMOs) (Marliere, 2009; Marliere *et al.*, 2011; Schmidt, 2010)

a) DNA-based biological circuits may be based on standard biological parts (but not necessarily).
b) The list is far from exhaustive – for example, with regard to complexity, in the more distant future we could also think of engineered tissues and organs, of altered multi-cellular and higher organisms, or even of entirely synthetic ecosystems composed of fully synthetic organisms.

The technical basis of SB is provided by DNA synthesis. Over recent years, scientists have been able to chemically synthesize ever longer stretches of DNA, from complete viruses genomes to whole bacterial genomes and synthetic (mini)chromosomes (Cello, Paul, and Wimmer, 2002; Tumpey *et al.*, 2005; (Carlson *et al.*, 2007; Macnab and Whitehouse, 2009; Wimmer *et al.*, 2009; Gibson *et al.*, 2010).

Advanced genetic constructs are used in metabolic engineering, however it remains a tedious process that triggered the interest in rationalization and reduction in design complexity. One attempt by SB is the creation and use of so-called standard biological parts. The idea is to have a toolbox of well-characterized, prefabricated, standardized and modularized genetic components (i.e., sequences of DNA) for engineering biological systems. Such standard biological parts could be freely combined and built into larger "devices" that fulfill certain defined functions. The hope is that these bioparts, will work as designed when re-introduced into "systems" or larger DNA-based biological circuits. So far, however, few properly working devices have been created, and parts characterization remains a challenge. To date, typical applications have been rather simple, for example, a chemical oscillator, banana-scent producing *Escherichia coli*, light-sensitive bacteria and so on (Elowitz and Leibler, 2000; Levskaya *et al.*, 2005; Greber and Fussenegger, 2007; Stricker *et al.*, 2008; Carlson, 2010).

One of the major problems in the design of biological systems is the lack of understanding of life's enormous complexity. The minimal genome research tries to reduce this complexity in order to define a minimal form of life (Itaya, 1995; Mushegian, 1999; Koonin, 2000). The goal is to create a simple cell (or platform or chassis) with the smallest possible genome still able to survive under laboratory conditions (Danchin, 2009; Deplazes and Huppenbauer, 2009). The minimal genome also has the advantage of serving as a cellular platform for engineered biological circuits, reducing possible interactions between the chassis and the crafted genetic circuits introduced.

The minimal genome is contrasted by the bottom-up approach to construct so-called protocells or synthetic cells. This is another attempt at constructing minimal versions of life, that is, synthetic cell-like vesicles, but in this case they are assembled bottom-up from nonliving chemical components (Rasmussen, 2009; Walde, 2010). The starting point is the construction of membrane-like structures to separate the inner from the outer world as well as those that enable a simple metabolism and procreation by fission, that is, the emergence of "daughter cells" (Szostak, Bartel, and Luisi, 2001; Hanczyc and Szostak, 2004). The SB subfields described above apply biochemistry in a similar way to natural life forms. For practical reasons, however, metabolic engineering aims at separating the newly introduced, constructed pathways from the naturally occurring ones; it seeks to introduce a certain degree of "orthogonality" while using the same or very similar building blocks.

Alternatively, the biochemistry of life could be entirely different, giving rise to xenobiology. Fully orthogonal systems are biochemical pathways that cannot interfere with naturally occurring ones at all (Schmidt, 2010). They arise from chemical synthetic biology, where the very basics of life biochemistry are changed in order to create biological systems that are truly different, both in metabolism and on the genetic information level. Examples include altered or non-naturally occurring bases within the DNA and comprise the idea of entirely different genetic information storage molecules, so-called xeno-nucleic acids (XNA) that cannot interact with naturally occurring DNA (Benner and Sismour, 2005; Yang *et al.*, 2006; Yang *et al.*, 2007; Herdewijn and Marliere, 2009; Marliere, 2009). Another example is the use of non-natural building blocks such as non-canonical amino acids. More visionary ideas even toy with the idea of replacing carbon with silicon in essential biomolecules. Other visions encompass possible life forms not only using non-natural elements but also architectures entirely different from the genome–ribosome–protein architecture of "life as we know it" (Schmidt, 2010).

5.2.1.2 **Synthetic Biology versus Genetic Engineering**

The distinction between SB and genetic engineering is not always very clear. For example, many forms of DNA-based biological circuits do in fact look quite similar to traditional forms of metabolic engineering or an advanced form of genetic engineering. The construction of protocells or xenobiological systems, in contrast, are typically lacking from the portfolio of genetic engineering techniques. It seems that some forms of SB are more closely related to genetic engineering than others and that the distinction is not a clear line but rather a fuzzy transit (Figure 5.1).

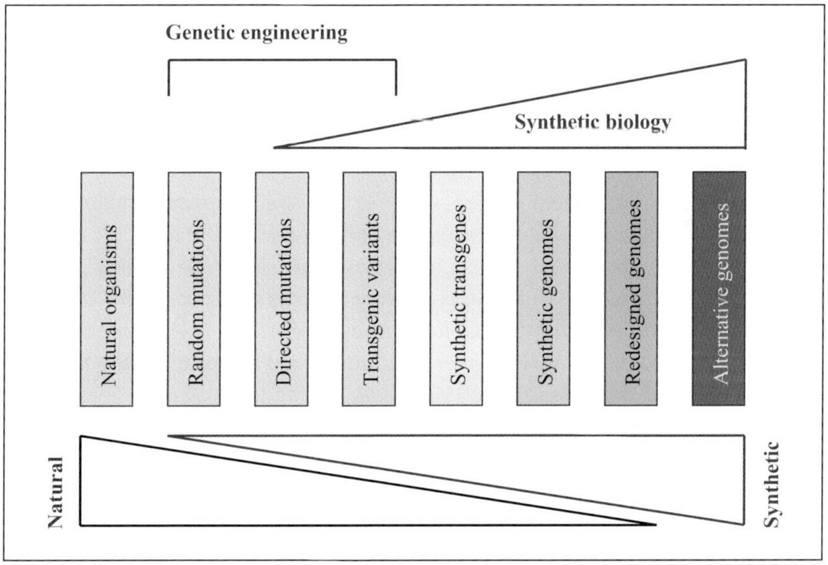

Figure 5.1 Stages of transition between naturally occurring organisms and wholly synthetic microbes. There is considerable overlap between genetic engineering and SB. Based on (de Lorenzo, 2010).

5.2.2
Existing Regulations

5.2.2.1 European Union

In Europe, SB is a considered a "new research field within which scientists and engineers seek to modify existing organisms by designing and synthesizing artificial genes or proteins, metabolic or developmental pathways and complete biological systems in order to understand the basic molecular mechanisms of biological organisms and to perform new and useful functions" (European Commission, 2009d). The European Commission (EC), already through Framework Programme 6 (FP6), supported research to investigate the capacity of doing SB within the European Union and to set up a strategic research policy. FP6 funded 18 projects ranging from energy to healthcare as reviewed by the NEST PATHFINDER initiative (European Commission, 2007b). The European Academies Science Advisory Council (EASAC) investigated policy questions surrounding the further funding for SB within FP7, releasing their reports in 2010 and 2011 (EASAC, 2010, 2011).

General Provisions Since the current state of the art of SB is still linked to genetic engineering approaches, the regulations relevant for SB within the European Union (EU) are mainly those for GMOs, based on the Directives and their amendments introduced since 1990 (European Commission, 1990a). Up to now neither the European Community nor any EU member state has issued specific legislation on SB. The national legal frameworks of EU member states, respectively, again are the result of adopting the EU legislation into their national system, particularly those on GMOs, medicine, biosafety, chemicals, data protection and patents. Global provisions from the World Health Organization (WHO) for biosafety standards and from the World Trade Organization (WTO) for supplements for international trading are taken account of within this regulatory body (European Commission, 2009c).

The European Parliament issued a couple of legislations to implement the Cartagena Protocol on Biosafety, focussing on risk assessment and risk management for the environmental release of GMOs (European Commission, 2001a), transboundary movements of GMOs (European Commission, 2003c) and contained use of GMOs (European Commission, 1998). The Laboratory Biosafety Manual published by the WHO serves as an important template for the biosafety standard for the EU regulations (WHO, 2004).

Presently, most of the activities involved in SB fall under the scope of regulations to the contained use of genetically modified microorganisms, or GMMs (European Commission, 1998). For all activities, the principles of good microbiological practice and good occupational safety and hygiene apply in accordance with the relevant community legislation. For research implying possible risks (to human health or the environment), it is required to inform the competent authority. For research involving higher risks, authorities must carry out prior risk assessment.

Environmental applications of SB are regulated according to the regulations for the intentional release of GMOs to the environment: preventive actions should be taken and the public is to be informed (European Commission, 2001a). Regarding commercial applications involving SB, they are covered by existing sectorial regulations. For example, there are regulations for medical products (European Commission, 2001b, 2003a, 2003b, 2004b), medical devices (European Commission, 1990b, 1993), new therapies (European Commission, 2001b, 2002b, 2004a, 2007a), cosmetic products (European Commission, 1976) and chemicals (European Commission, 2006a, 2006b). Data protection is regulated specifically (European Commission, 1995, 2002a), as is safety and health protection of workers from risks related to exposure to biological agents at work (European Commission, 2000b).

There are also legislations on notices the authority on the outbreak of infections in plants and animals (European Commission, 1982, 2000a). There are still, however, a few issues on the regulation on SB products that are not yet fully addressed, as listed by the EGE report, patenting and open access (European Commission, 2009c).

Dual Use and Biosecurity A major topic in the debate over regulatory issues around SB is the problem of how to treat dual use[81] applications and biosecurity. Regarding biosecurity, the Australia Group Guidelines[82] (BTWC, 1975) have been adapted on an EU level by Council Regulation (EC) No. 1334/2000 to provide guidance for the control of dual use items and technology, which also serves as a regulatory framework for biosecurity (European Commission, 2000c). The EU export control regime is governed by Regulation 428/2009 (European Commission, 2009a). The list of controlled dual use items is set out in itsAnnex I.

Items not listed in that Annex I may also be subject to export controls under certain conditions specified in the Regulation. Under the EU regime, controlled items may not leave the EU customs territory without export authorization. Additional restrictions are in place concerning the provision of brokering services with regard to dual use items and concerning the transit of such items through the EU. Taken together, there are four types of export authorization:

- **Community General Export Authorisations (CGEA)** cover exports of most controlled items to the United States, Canada, Japan, Australia, New Zealand, Switzerland, Norway. A proposal exists to create new CGEAs to simplify the current system with regard to exports of certain items to low-risk destinations;
- **National general export authorizations (NGAs)** may be issued by individual EU countries provided that:

81) Dual use goods are products and technologies normally used for civilian purposes but which may have military applications.

82) See: http://www.australiagroup.net/en/index.html

  - They do not conflict with existing CGEAs;
  - They do not cover any of the items listed in part 2 of Annex II to Regulation 428/2009;
  - France, Germany, Greece, Italy, Sweden, the Netherlands and the United Kingdom currently have these authorizations;
- **Global authorizations** are granted by individual EU countries to one exporter and cover one or more items to one or more countries/end users;
- **Individual licenses** are granted by individual EU countries to one exporter and cover exports to one end user.

A particular problem is considered to be the export of synthesized DNA sequences. Although not mentioned explicitly, Regulation 428/2009 also includes companies performing DNA synthesis and international shipping thereof. It states that (emphasis added):

> The Community General Export Authorisation may not be used if:
>
> - The exporter has been informed by the competent authorities of the Member State in which he is established that the items in question are or may be intended, in their entirety or in part, for use in connection with the development, production, handling, operation, maintenance, storage, detection, identification or dissemination of chemical, biological or nuclear weapons or other nuclear explosive devices or the development, production, maintenance or storage of missiles capable of delivering such weapons, or **if the exporter is aware** that the items in question are intended for such use;
> - The exporter has been informed by the competent authorities of the Member State in which he is established that the items in question are or may be intended for a military end use as defined in Article 4(2) of this Regulation in a country subject to an arms embargo decided by a common position or joint action adopted by the Council or a decision of the OSCE or an arms embargo imposed by a binding resolution of the Security Council of the United Nations, or **if the exporter is aware** that the items in question are intended for the above mentioned uses;
> - The relevant items are exported to a **customs free zone or free warehouse** which is located in a **destination covered by this authorization**.

This means that in the case of DNA synthesis, the burden falls back to the company and they have to check whether the product (i.e., the DNA sequence) ordered by a third-country costumer may be intended for biological warfare.

Within the EU dual use items may be traded freely, except for those listed in Annex IV to Regulation 428/2009, which are subject to prior authorization. On top of that, national authorities may require export controls on unlisted dual-use items (see Articles 4 and 8 of Regulation 428/2009). Exporters should therefore refer to their relevant national rules and check the situation with regard to their specific transactions.

Individual EU countries may keep in place certain specific national rules. Such rules can apply to additional items to be controlled (Articles 4 and 8). They can require goods to be checked at specific border points (Article 17). They can introduce additional checks inside the EU (Article 11).[83]

The EU list of controlled items is based on control lists adopted by international export control regimes – the Australia Group (AG), the Nuclear Suppliers Group (NSG), the Wassenaar Arrangement and the Missile Technology Control Regime (MTCR).

Apart from these international agreements, the European Community has developed the EU Counter-Terrorism Strategy adopted in 2005 (European Commission, 2005). This strategy covers four strands of work (of which the EC plays a role in the PREVENT and the PROTECT strand by providing assistance to the member states):

- PREVENT people from turning to terrorism and stop future generations of terrorists from emerging;
- PROTECT citizens and critical infrastructure by reducing vulnerabilities against attacks;
- PURSUE and investigate terrorists, impede planning, travel and communications, cut off access to funding and materials and bring terrorists to justice;
- RESPOND in a coordinated way by preparing for the management and minimization of the consequences of a terrorist attack, improving capacities to deal with the aftermath and taking into account the needs of victims.

Advisory Bodies Three safety-related scientific committees provide scientific advice to the European Commission:

- The Scientific Committee on Consumer Safety (SCCS);
- The Scientific Committee on Health and Environmental Risks (SCHER);
- The Scientific Committee on Emerging and Newly Identified Health Risks (SCENIHR).

The SCCS provides opinions on the basis of a specific set of data (dossier) submitted by an applicant (e.g., Industry, Member State authorities, etc) in order to satisfy specific regulatory requirements (e.g., Directive 76/768/EEC, Directive 2009/48/EC, etc.). To date, there is no opinion about product developed by SB approaches issued by SCCS (SCCS).

The SCHER provides opinions on health and environmental risks related to pollutants in the natural setting. It also assesses "the biological and physical factors or changing physical conditions that may have a negative impact on health

83) See: http://ec.europa.eu/trade/creating-opportunities/trade-topics/dual-use/

and the environment (e.g., in relation to air quality, waters, waste and soils)." In addition, it provides "opinions on life cycle environmental assessment and addresses health and safety issues related to the toxicity and eco-toxicity of biocides." It will publish its opinions by the end of the assessment and statements on specific topic (SCHER).

The SCENIHR "provides opinions on emerging or newly identified health and environmental risks and on broad, complex or multidisciplinary issues requiring a comprehensive assessment of risks to consumer safety or public health and related issues not covered by other Community risk assessment bodies." The potential activities overseen by this committee are currently covering those derived by new technologies such as nanotechnologies (SCENIHR). The SB-related activities and products would therefore also be assessed by this committee.

Overview of current European regulations relevant to SB

- **Risk assessment and risk management:**
  Directive 2001/18/EC on the deliberate release into the environment of genetically modified organisms;
  Regulation (EC) 1946/2003 on trans-boundary movements of genetically modified organisms;
  Council Directives 2009/41/EC and 98/81/EC amending Directive 90/219/EEC on the contained use of genetically modified microorganisms.
- **Biological risks:**
  Council Directive 82/894/EEC;
  Council Directive 2000/29/EC of May 8, 2000;
  EC Council Regulation No 1334/2000.
- **Occupational health:**
  Directive 2000/54/EC on safety and health for workers exposed to biological agents at work.
- **New medicinal products:**
  Regulation (EC) No 726/2004;
  Directive 2001/83/EC;
  Directive 2003/94/EC;
  Directive 2003/63/EC).
- **Medical devices:**
  Directive 93/42/EEC;
  Directive 90/385/EEC.
- **Gene therapy, cell therapy and tissue engineering:**
  Regulation (EC) No 1394/2007 amending Directive 2001/83/EC;
  Regulation (EC) No 726/2004;
  Directive 2001/83/EC;
  Directive 2004/23/EC;
  Directive 2002/98/EC.

Overview of current European regulations relevant to SB (*Continued*)

- **Clinical trials:**
  Regulation (EC) 2001/20 amended in 2003 and 2005.
- **Cosmetic products:**
  Directive 1976/768/EC.
- **Data protection:**
  Directive on the processing of personal data and the protection of privacy in the electronic communications sector.
- **Chemicals:**
  REACH, the European Community Regulation on chemicals and their safe use (EC 1907/2006). It deals with the **R**egistration, **E**valuation, **A**uthorization and Restriction of **Ch**emical substances. The law entered into force on June 1, 2007.
- **Patents:**
  Directive 98/44/EC.
- **Dual use export control:**
  Regulation 428/2009.

Global provisions:

- **WHO biosafety standards;**
- **The Cartagena Protocol; World Trade Organization (WTO) agreements;**
- **Trade-Related Aspects of Intellectual Property Rights (TRIPS)** on the limits of safety regulation measures;
- **Convention on the Prohibition ofthe Development, Production and Stockpiling of Bacteriological (Biological) and Toxin Weapons and on their Destruction;**
- **UN Security Council Resolution 1540.**

5.2.2.2 Examples of National Regulations

National regulations of EU member states reflect the decisions and regulations made in the European Commission. So national regulations have to comply with the European regulatory system, but may add additional laws and regulations on top of that (as long as they don't violate the EC regulations and laws). Scientists, industry and even do-it-yourself biological amateurs thus have to consider both national and EC laws and regulations when working with SB, genetically engineered organisms or chemicals. We chose four examples, two small countries (one EU member: Austria; one non-EU member: Switzerland) and two large countries (Germany, UK).

National authorities have compiled lists of donor and receptor organisms that already have undergone risk assessment. These lists are very similar, but only apply in the national context, such as Chapter 3, in the Austrian "Book on genetic engineering", the "Lists of organisms" in the German law, the "Categorization of biological agents according to hazard and categories of containment" in the United Kingdom or the "Annexe II.1–II.5.k" by the French Commission de Génie Génétique. These lists reflect the different modus of expert advice in various member countries rather than substantial differences in national risk assessment results.

5.2.2.3 Austria

Introduction In Austria, the status of SB is somewhat unclear as there are very few scientific groups explicitly performing synthetic biology work, and no special regulation has been issued so far. Neither is there a national funding scheme for SB, so any pertaining activity falls under the term of (general) biotechnology or, rather, genetic engineering, which is extensively regulated, however. Regulatory activity has been focussing the problem of GMO releases in the light of a lack of public esteem for agricultural biotechnology. Rules for the contained use of GMOs, in contrast, and thus the most important regulations for current SB applications, do not put up higher hurdles as in any other European member state.

General Provisions In addition to the European regulations, the Austrian jurisdiction has four legal categories dealing with genetic engineering and its products: (i) the Austrian Genetic Engineering Act, (ii) National Acts, (iii) Province Laws to regulate coexistence (of GM and nonGM plants) and (iv) other regulations.

The Austrian Gentechnikgesetz (Genetic Engineering Act) has been issued in 1995. It controls the work with GMOs in contained systems (laboratory, production facility) as well as deliberate releases into the environment and the placing on the market of GMOs, gene analysis and gene therapy in humans. The aim of this law is to protect human health and the environment from direct and indirect harmful effects of GMOs, but originally also to prevent "social unsustainability" of GM products; a stipulation that never had any practical importance, though. The original act has been changed several times, especially in order to incorporate the relevant EC guidelines 2001/18/EC and 98/81/EC. (Umweltbundesamt, 2008). In order to work with GMOs in Austria, be it in contained systems or for environmental release, the competent Austrian authority has to grant its consent. The notifier exclusively is authorized to use a GMO or GMOs in compliance with the conditions stipulated in the written consent. (Umweltbundesamt, 2004).

Apart from the Genetic Engineering Act, there are a number of national acts with a more specific focus (Umweltbundesamt, 2008):

- The **Systems Act** regulates the use of GMOs (microorganisms) in contained facilities (laboratories, production facilities, glass houses etc) and contains detailed biosafety regulations (e.g., use of biosafety levels 1–4 for different organisms).
- The **Environmental Release Act** specifies the form and content of release applications of GMOs and their bringing onto market. It provides applicants with all relevant information and describes the safety assessments that have to be completed before the application (including an environmental impact assessment) and provides details about monitoring activities (once the GMO has been released into the environment).

- The **Public Hearing Act** details the participation of the general public in environmental release applications. Each application has to be made public and anybody can comment on the application (for a certain period of time).
- The **Labeling Act** refers mainly to the EC guidelines on labeling of food and feed products and deals with labeling of products that are neither food nor feed.
- The **GMO Registry Act** describes all aspects that need to be completed in the so-called Austrian GMO Registry. The Registry contains all GMO products that have been granted permission for deliberate release or for marketing.
- The **GM Seed Act** was issued in 2011 after problems of "contamination" of nonGMO seeds with GMO seeds. NonGM seeds must not contain more than 0.1% of GMO seeds (otherwise they are considered genetically modified).
- The **Seed Growing Area Act** was put in place in 2005 and allows the prohibition of GM crop deployment in regions that produce nonGMO seeds (for further sale to farmers).
- The **Simplification Act** allows for a quicker bureaucratic evaluation of applications that are follow-ups of already approved varieties.
- The **GMO Feed Threshold Act** regulates the use of GM products in feed and limits the "contamination" of nonGM feed with GMO-derived products by 1%.
- The **Genetic Engineering Waste Act** handles thresholds for waste emissions related to the use of GMOs in contained facilities.
- A number of specific **Prohibition Acts** impede the import and/or sale of certain GM crops.

In addition, the **Province Acts** issued by the nine provincial governments deal with questions of coexistence of traditional/organic crops and GM crops. The Acts intend to protect traditional/organic crop production from "contamination" of GM crops (e.g., pollen). A general ban on GM crops was initially issued by the government of Upper Austria but was rejected by the European Commission and the Court of Justice of the European Union. GM-free regions can only be established if there is an ecological reasoning behind (e.g., protecting the natural integrity of a national park).

Other regulations include, for example, the Codex Alimentarius Austriacus, which defines the conditions under which GM-free food may be produced, or the Austrian Book on Genetic Engineering (Gentechnikbuch), which documents the R&D state of the art in genetic engineering.[84)]

Advisory Bodies A two-tiered system of advisory bodies is in place, with a general Gentechnikkommission, which is, in practice, rather uninfluential. More powerful

84) For more information see also: http://www.bmwf.gv.at/nc/startseite/forschung/national/forschungsrecht/gentechnik/?sword_list[0]=gentechnik and http://www.umweltbundesamt.at/umweltsituation/gentechnik/

are the scientific panels on applications in closed systems, on GMO releases and on medical biotechnology. They advice the competent authorities, regarding the scientific aspects of applications for permissions and other issues.

Austrian regulation for genetically modified organisms	Specific law (Federal Legal Gazette)
Genetic engineering act (Austrian Gentechnikgesetz)	BGBl. Nr. 510/1994
First change of the genetic engineering act	BGBl. I Nr. 73/1998
Second change of the genetic engineering act	BGBl. I Nr. 94/2002
Third change of the genetic engineering act	BGBl. I Nr. 126/2004
Fourth change of the genetic engineering act	BGBl. I Nr. 127/2005
Fifth change of the genetic engineering act	BGBl. I Nr. 13/2006
System act	BGBl. Nr. 431/2002
Environmental release act	BGBl. II Nr. 260/2005
Public hearing act	BGBl. Nr. 61/1997, BGBl. II Nr. 164/1998
GMO labeling act	BGBl. II Nr. 5/2006
GM registry act	BGBl. II Nr. 141/2006
GM seed act	BGBl. II Nr. 478/2001
Seed growing area act	BGBl. II Nr. 128/2005
Genetic engineering waste act	BGBl. II Nr. 350/1997
Act on biological working substances	BGBl. II Nr. 237/1998
Prohibition act GM Maiz MON 810	BGBl. II Nr. 181/2008
Prohibition act GM Maiz T25	BGBl. II Nr. 180/2008
Prohibition act GM Rapeseed GT73	BGBl. II Nr. 157/2006, BGBl. II Nr. 307/2010
Prohibition act GM Rapeseeds Ms8xRf3	BGBl. II Nr. 246/2008, BGBl. II Nr. 305/2010
Prohibition act GM Maiz MON 863	BGBl. II Nr. 257/2008, BGBl. II Nr. 306/2010
Prohibition act GM Potato EH92-527-1	BGBl. II Nr. 125/2010

5.2.2.4 **Germany**

Introduction Similar to the situation in the European Commission and other European countries, no law or regulation has been passed in Germany that deals with specifically with SB. Practically all SB activities currently fall under the German law regulating genetic engineering[85] initially passed in 1990. Since then

85) See full text of the law at: http://www.gesetze-im-internet.de/gentg/index.html

it has undergone changes and improvements (in 2004, 2006 and 2008), mostly in response to new pieces of European legislation.[86)]

General Provisions Covering all types of organisms except humans, the law regulates safety requirements for genetic engineering in contained facilities, the administrative process of granting permissions for deliberate releases of GMOs into the environment, the placing on the market of GMO products, liabilities and penalties in seven parts:

1) General provisions;
2) Genetic engineering in genetic engineering facilities;
3) Release and placing on the market;
4) Common provisions;
5) Liability provisions;
6) Penalty provisions;
7) Transition and final provisions.

The law has been designed to allow for a positive development and application of genetic engineering. It aims at supporting research and development in Germany without jeopardizing the safety of humans or the environment and without restricting the freedom of choice (e.g., when buying food products) for consumers. Another important aspect is the provision for a co-existence of different agricultural production systems (i.e., GM and nonGM).

The latest changes in the law in 2008 included improvements for the R&D institutions to make sure that Germany does not fall behind and become dependent on the leading countries in science and technology.

Similar to Austria, the majority of regulatory changes in recent years focused on agricultural biotechnology and the public resistance encountered. Among the many changes, many deal with the co-existence of GM and nonGM crops. For example, according to Section 16a, a public registry of locations with GM crops was set up (also available on the internet), allowing the public to go searching for GM crop fields in Germany (see Figure 5.2). This disclosure has often led to environmental groups destroying the fields.

In addition, labeling of GM products is mandatory according to Section 17b. The law also contains an unusual specification (Section 33) of the maximum penalty of € 85 Mio for damage caused by genetic engineering.

Advisory Bodies In Section 4, the German law defines the "Central Commission for Biological Safety" (Zentrale Kommission für die Biologische Sicherheit; ZKBS), an independent expert commission consisting of scientists from different fields, but also representatives from industry, workers' unions, agriculture, consumer and environmental protection organizations and science funding bodies. Their goal is to clarify open questions and challenges when applying the law.

In terms of DNA synthesis, dual use and biosecurity, Germany is part of the Australia Group and the Wassenaar Arrangement. It follows the EC regulation on

86) See: http://www.bmelv.de/SharedDocs/Standardartikel/Landwirtschaft/Pflanze/GrueneGentechnik/Gentechnikrecht.html

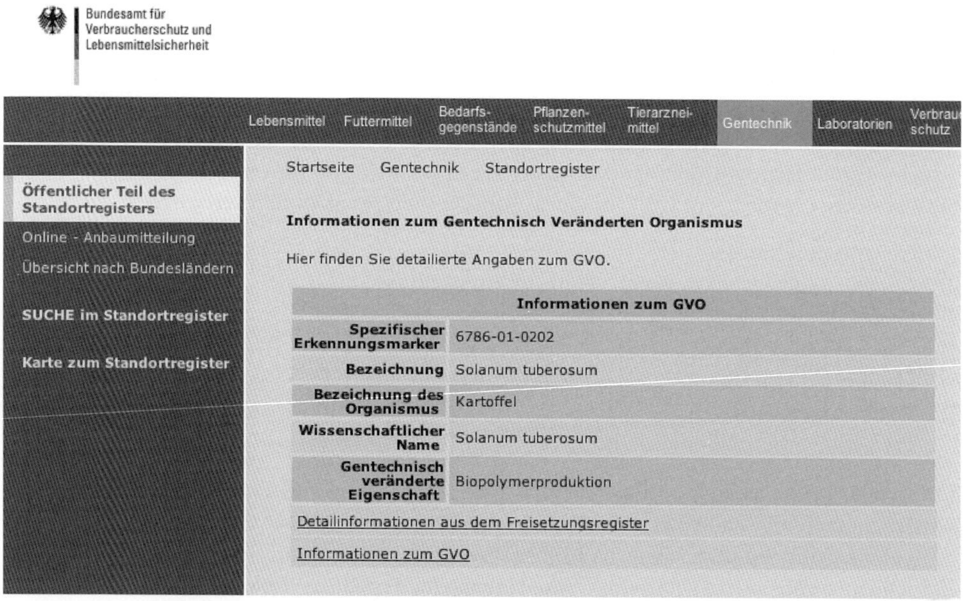

Figure 5.2 Publicly available registry of GM crop field release in Germany. Source: http://apps2.bvl.bund.de/stareg_web/showgvo.do?gvoId=346.

dual use (428/2009) and has also translated its content into national law, to the so-called Law controlling weapons of war (Kriegswaffenkontrollgesetz[87]). In Section 18 it explicitly forbids the development, production, commerce, import, export or transport within Germany of any bioweapon.

The Federal Office of Economics and Export Control (BAFA, Bundesamt für Wirtschaft und Exportkontrolle) keeps an eye on this and other dual use regulations and their compliance. For example, each German company (such as a DNA synthesis company) has to name a person who is responsible for export issues (Ausfuhrverantwortlichen) who also has to have a senior management position in the company. This person is responsible to report to the Federal Office in case of any dual use issues.[88] In practice this means that, for example, the CEO of a German DNA synthesis company is responsible for making sure that the company does not export synthetic DNA that might be used for bioterrorism. In practice German DNA synthesis companies have formed two international industry associations that helps them avoid the inadvertent production or export of a dual use product. The International Association Synthetic Biology (IA-SB) issued a Code of Conduct for Best Practices[89] in Gene Synthesis, and the International Gene Syn-

87) See: http://www.gesetze-im-internet.de/bundesrecht/krwaffkontrg/gesamt.pdf

88) See: http://www.bafa.de/ausfuhrkontrolle/de/vorschriften/zuverlaessigkeit_ausfuhrverantwortlicher/index.html

89) See: http://www.ia-sb.eu/go/synthetic-biology/synthetic-biology/code-of-conduct-for-best-practices-in-gene-synthesis/

thesis Consortium (IGSC) came up with a Harmonized Screening Protocol.[90] Both groups carry out a twofold screening strategy: first they screen the gene sequence to determine whether it is relevant for dual use (i.e., resembles a sequence for a bioweapon substance, or helps with weaponization) and second they screen the customer (country, organization, individual). Based on this assessment they can decide whether or not the DNA synthesis order can be accepted.

5.2.2.5 **United Kingdom**

Introduction In the United Kingdom (UK), SB is defined as "the deliberate design of novel biological systems and organisms that draws on principles elucidated by biologists, chemists, physicists and engineers [. . .] in essence it is about redesigning life." (The Royal Society, 2008). The postnote report on SB (a short account for Members of Parliament by the Parliamentary Office of Science and Technology) already published in 2008 stated that: "synthetic biology is an immature but rapidly developing area involved with research into novel, engineered purpose-built biological parts, devices and organisms", and "The UK's synthetic biology community is small and funding for this type of multidisciplinary research falls within the remits of several research councils" (Bunn, 2008). It emphasizes that SB can construct novel genomes from standardized parts rather than merely transfer genes from a donor to a recipient cell like genetic modification does.

General Provisions Sharing many features with genetic modification, though SB is covered by the GM regulations in the UK such as the Genetically Modified Organisms (Contained Use) Regulations 2000 (with amendment regulations in 2002, 2005, 2010). This legislation provides guidance on the risk assessments of, as well as how to conduct activities involving, genetic modification (2000).

> The Genetically Modified Organisms (Contained Use) Regulations 2000:
>
> - Require risk assessment of activities involving genetically modified microorganisms and activities involving organisms other than microorganisms. All activities must be assessed for risk to humans and those involving GM microorganisms (GMMs) assessed for risk to the environment.
> - Introduce a classification system based on the risk of the activity independent of the purpose of the activity. The classification is based on the four levels of containment for microbial laboratories.
> - Require notification of all premises to the Health and Safety Executive (HSE) before they are used for genetic modification activities for the first time.
> - Require notification of individual activities of Class 2 (low risk) to Class 4 (high risk) to be notified to the competent authority (which HSE administers). Consents are issued for all Class 3 (medium risk) and

90) See: http://www.genesynthesisconsortium.org/Gene_Synthesis_Consortium/Harmonized_Screening_Protocol.html

Class 4 (high risk) activities. Class 1 (no or negligible risk) activities are non-notified, although they are open to scrutiny by HSE's specialist inspectors who enforce the Regulations. Activities involving GM animals and plants that are more hazardous to humans than the parental non-modified organism also require notification.

- Require fees payable for the notification of premises for first time use, class 2, 3 and 4 activity notifications and notified activities involving GM animals and plants.
- Require the maintenance of a public register of GM premises and certain activities.

The Department for Environment Food and Rural Affairs (DEFRA) is responsible for environmental releases of genetically modified organisms. Legislation covering environmental risks associated with GM plants and animals includes:

1) Section 108 (1) of the Environment Protection Act 1990 (Legislation UK, 1990);
2) The Genetically Modified Organisms (Risk assessment, Records and Exemptions) Regulations 1996 (Legislation UK, 1996);
3) The Genetically Modified Organisms (Deliberate Release and Risk Assessment–Amendment) Regulations 1997 (Legislation UK, 1997).

Advisory Bodies As the national independent watchdog for work-related health, safety and illness, the HSE is a member of the SB Policy Coordination Group. The Group under the auspices of the Royal Society is composed of representatives from academia (engineers, chemists, biologists and social scientists), government (HSE, Department of Health) and research funding agents (Environment Science Research Council and Wellcome Trust). This expert group is to advise on the application of GM regulations to SB, aiming at exchanging information on national and international departments, identifying gaps in current policy, minimizing duplication and promoting collaborations (The Academy of Medical Sciences and Engineering, 2007).

The Health and Safety Commission appointed the Scientific Advisory Committee on Genetic Modification (SACGM) as part of its formal advisory structure. In 2007, the SAGCM published a guidance paper, SACGM Compendium of Guidance (HSE, 2007b, 2007c), in consultation with the HSE, which reflects good practice but is not compulsory. For academic activities involving genetic modifications (such as SB), there are nevertheless certain legal requirements:

1) The risk has to be assessed and noticed.
2) The activity has to be classified (class 1–4, with equivalent containment levels).
3) Formal approval has to be given by either the Genetic Modification Safety Committee (GMSC) of the host institute, the Biological Safety Officer (BSO) or the Chair of the GMSC.

5.2.2.6 Switzerland

Introduction Switzerland, although not a member of the European Union, has been actively involved in European scientific funding and policymaking regarding SB; and Swiss regulations on science policy are consistent with European norms, and rules on SB are adapted from those on GMOs (Furger and Schweiz, 2007). SB is considered a relatively new field of research and a combination of molecular biology, chemistry, computer science and engineering. Essentially, SB is considered to imply research on organisms rebuilt or designed in a controlled manner for specific purposes (CH-ECNH, 2010).

General Provisions Under Swiss law issued by The Federal Authorities of the Swiss Confederation, SB-related techniques are considered to be "techniques for genetic modification" (sc | nat, 2006). SB research thus is subject to the regulations on "gene technology" and the federal law for environmental protection. Synthetic organisms and their use are regulated by the Gentechnikgesetz (GTG; CH-GTG, 2003). Environmental protection has to comply to the Umweltschutzgesetz (USG; CH-USG, 1983) and the safety law (CH-Sicherheit, 1996), the environmental release of GMOs to the Freisetzungsverordnung (FrSV; CH-FrSV, 2008). Research involving pathogenic organisms is regulated under the Epidemiengesetz (CH-Epidemiengesetz, 1970). For biosafety management, guidelines are provided by the regulations governing personal injury insurance (CH-UVG, 1981) and protection against hazards caused by microorganisms (CH-SAMV, 1999).

The main regulatory authorities for SB are the Bundesamt fuer Umwelt (BAFU), aka the Federal Office for the Environment (FOEN), and the Bundesamt fuer Gesundheit (BAG), aka the Federal Office of Public Health (FOPH). The FOEN is responsible for ensuring that natural resources are used in a sustainable manner, the public is protected against natural hazards, and the environment is protected from unacceptable adverse impacts. Biotechnology is one of the key research topics of FOEN and it provides regulations for activities in contained use, experimental release and marketing (of GMOs, pathogenic organisms and alien small invertebrates), international relations, research and legal bases (CH-FOEN, 2011). The FOPH is part of the Federal Department of Home Affairs and is responsible for public health and the development of a national health policy. It issues legal directives on consumer protection (particularly in relation to food, chemicals, therapeutic products, cosmetics and utility goods) and supervises their implementation. It is responsible for monitoring transmissible diseases and for radiological protection and issues the necessary regulations (Ch-FOPH, 2011).

Dual Use and Biosecurity Similar to the EC, Switzerland has an export control regulation for dual use goods – Federal Act on the Control of Dual Use Goods and Specific Military Goods, (SECO, 2004, 2005). Among other potential exporters, it is relevant to DNA synthesis companies. The act states (emphasis added): "The planned export of goods that is not subject to the license requirements under Article 3 must be reported to SECO[91] in writing if:

91) Swiss State Secretary of Economic Affairs, see: www.seco.admin.ch/

a) the **exporter knows that the goods are intended or could be intended wholly or partly for the development, production or use** of nuclear, **biological** or chemical **weapons** (NBC weapons) or of delivery systems for the use of NBC weapons, or for the construction of facilities for NBC weapons or their delivery systems"

Hence, the legal situation in this respect is very similar to that in the EU.

Advisory Bodies Two boards provide advice to the authorities from an expert perspective. The Eidgenössische Fachkommission fuer biologische Sicherheit (EFBS), aka the Swiss Expert Commission for Biological Safety (SECB) provides opinion from a biosafety perspective (CH-SECB, 2011). The SECB advises the Federal Council and competent authorities on the drafting and the enforcement of laws, ordinances, guidelines and recommendations. It issues statements on license applications and publishes recommendations on safety measures for studies using genetically modified or pathogenic organisms.

From an ethical perspective, the Eidgenössische Ethikkommission fuer die Biotechnologie im Ausserhumanbereich (EKAH), aka the Federal Ethics Commission on Non-Human Biotechnology (ECNH; CH-ECNH, 2011) advises the Federal Council and the competent authorities on legislation and enforcement in biotechnology and gene technology. The committee consists of 12 independent experts from various fields and aims at ensuring a responsible approach to SB.

Swiss legislation relevant for SB

Federal Constitution:

- Federal Constitution of the Swiss Confederation, Art. 74: Protection of the Environment;
- Federal Constitution of the Swiss Confederation, Art. 120: Gene Technology in the Non-Human Field.

Acts and Ordinances

Environment:

- Gene Technology Act (GTA);
- Federal Act relating to Non Human Gene Technology (Gene Technology Act; GTA);
- Environmental Protection Act (EPA)

Release Ordinance (RO);

- Ordinance on the contained use of organisms (ESV);
- Ordinance on the contained use of organisms (Containment Ordinance, CO);
- Ordinance on the Environmental Impact Assessment (UVPV);
- Ordinance on Protection against Major Accidents (StFV);
- Ordinance on the Swiss Expert Committee for Biosafety.

Swiss legislation relevant for SB *(Continued)*

Employees:

- Federal law on accident insurance (UVG);
- Ordinance on occupational safety in biotechnology (SAMV);
- Ordinance on the prevention of professional accidents and illnesses (VUV).

Food:

- Federal law on food (LMG);
- Ordinance on food (LGV);
- Ordinance on genetically modified food (VGVL).

Agriculture:

- Landwirtschaftsgesetz (LwG);
- Ordinance on seeds;
- Ordinance on Plant Protection Products (PSMV);
- Dünger-Verordnung (DüV);
- Ordinance on Plant Protection (PSV);
- Feedstuff Book Ordinance (FMBV);
- Ordinance on GMO feed list.

Export control regulations:

- Federal Act on the Control of Dual-Use Goods and of Specific Military Goods;
- Ordinance on the Control of Dual-Use Goods and Specific Military Goods.

Others:

- Federal law on epidemics;
- Federal law on epizooites;
- Federal law on animal protection;
- Federal law on patents;
- Conventions and international agreements ratified by Switzerland;
- Convention on Biological Diversity;
- Cartagena Protocol on Biosafety.

5.2.3
Options for Adapting and Improving Regulations

SB develops in the context of existing biosafety and biosecurity rules, regulations and laws mainly put into place to deal with genetic engineering. Departing from this regulatory context, however, experts pointed out the need to adapt safety rules to upcoming developments in SB (Bugl *et al.*, 2007; Garfinkel *et al.*, 2007; Kelle, 2009; Schmidt, 2009; Schmidt, 2010; Schmidt *et al.*, 2009; National Institutes of Health, 2010; Schmidt and Pei, 2011). Sooner or later, therefore, the current set of regulations and provisions could turn out to be insufficient to fully deal with more advanced forms of SB (de Lorenzo, 2010). Since there is no defined threshold between genetic engineering and synthetic biology but a gradual increase in "artificialness," any boundary would have to be deliberately set. Hence, defining SB and subjecting the so-defined field to regulatory conditions different from biotechnology at large would demand a bold political decision.

The underlying reason for the concern is that, with health and occupational safety standards, problems with anticipating risk from SB might arise. Established risk mitigation measures are built upon classifications of naturally occurring organisms. Since products of synthetic biology are intended to be entirely novel, it may become questionable whether these classification schemes fully apply. Therefore, the question arises whether and how fundamentally novel objects derived from SB can be assessed applying the traditional approaches of comparing unknown living organisms with known ones (Schmidt, 2009). Risk assessment might have to change because:

> . . . safety risks for genetically modified organisms might not be best judged from the behaviour of the parent organism, once modification is pursued at the 'deep' systemic level that synthetic biology should enable. (IRGC, 2010)

The question whether the existing body of biotechnology regulation, including the provisions for risk assessment would suffice to prevent potential risks from SB to materialise has spurred considerable activity worldwide. For example, the United States President's Committee on Bioethics stated that:

> Because of the difficulty of risk analysis in the face of uncertainty – particularly for low-probability, potentially high-impact events in an emerging field – ongoing assessments will be needed as the field progresses. Regulatory processes should be evaluated and updated, as needed, to ensure that regulators have adequate information. . . . It [Executive Office of the President] should also identify any gaps in current risk assessment practices related to field release of synthetic organisms. And, recognizing that international coordination is essential for safety and security, the government should act to ensure ongoing dialogue about emerging technologies such as synthetic biology (Presidential Commission for the Study of Bioethical Issues, 2010).

According to the European Group on Ethics (EGE), most current activities in SB and its foreseen fields of application are already regulated at the European Union level (European Commission, 2009c). The main argument supporting this opinion is a difficulty to distinguish early stages of SB from enhanced stages of genetic engineering. As a consequence, the regulatory context for genetic engineering also applies to the initial (current) developments of SB. So first SB products have to comply with the existing regulations. In addition, the report by the EGE suggested that: *"the EC initiates a survey on relevant risk assessment procedures in the EU and identify possible gaps in the current regulations and how identified gaps are to be filled; and that a Code of Conduct for research in synthetic biology should be prepared by the EC"* (European Commission, 2009b). Further, "*. . . EU Biosafety standards for Synthetic Biology products [should be] adopted . . .*".

This call elicited different reactions. For example, SCENIHR suggested that four steps should be considered while assessing the risks posed by SB (European Commission, 2010):

1) Characterization of the relevant physical and chemical properties of each product or process, along with its biological properties;
2) Assessment of the potential exposure of humans, animals and the environment under expected and misuse conditions;
3) Examination of the hazardous properties;
4) Estimation of the risk.

Not only on the European Community level, but even more so on the national level, a proposal such as that from SCENHIR met with multiple preceding and on-going pertaining activities. Without claiming comprehensiveness, we would like to present some voices from within the advisory bodies of some European countries.

Among those countries that have dealt with policy challenges from SB, the United Kingdom has been particularly active in elucidating the implications for society and especially for risk and its regulation. Several initiatives have been taken to clear the grounds for an appropriate future policy on SB. To shape the focus, the Synthetic Biology Policy Coordination Group of the Royal Society issued a "call for views" (Royal Society, 2007). In their report, the group emphasized that existing regulations such as those relating to biosafety, biosecurity and the release of GMOs into the environment should be explored whether they still are adequate and match likely developments and future applications of SB. Under the header of "responsible development", a dialog among stakeholders on social and ethical issues in an early stage of technological development and an independent assessment would be needed to avoid hyping potential benefits and risks. More recently, the Royal Academy of Engineering conducted a survey of public perceptions on SB in the United Kingdom (Royal Academy of Engineering, 2009), showing that the public wanted governmental regulations on SB as long as they do not impede development. To provide accessible and accurate information about SB, the EPSRC founded a "Center for SB and Innovation" dedicated to foster public engagement.

From March 2007 on, the HSE released several reports related to SB such as the Horizon Scanning Intelligence Group Short Report (HSE, 2007a), where they emphasized the relevance to issues of health and safety, especially the possible implications of synthetic microorganisms posing risks to human health and the environment due to their ability to reproduce and evolve. The risks associated with traditional GM approaches can, accordingly, be estimated based on the risks posed by the unmodified donor and acceptor organisms. In contrast, SB-derived GMOs require a higher containment measure due to the facts that SB activities are complex, indefinable and difficult to anticipate precisely. The HSE recommended that the regulation of SB should follow that on GMOs, but they also emphasized the need for active monitoring. In addition, they pointed out that many researchers working in SB were engineers, who supposedly command limited knowledge in genetic modification. Accordingly, they might also be less likely to fully understand the regulatory system and its demands.

Apart from the regulations on GMOs, nanotechnology regulation has been influential as an example to follow. Not only is nanotechnology a tool to accelerate the advancement of SB; in the United Kingdom it seems to also have been considered a blueprint for how to deal with unexpected risks from synthetic biology. Regarding regulatory policy development, both emerging technologies are considered to provoke identical issues in debates such as risk governance, adequacy of the regulatory framework, public perception, risk/benefit judgments and ethical, legal and societal issues (ELSI). In 2009, DEFRA released a report to call for a code of conduct in nanotechnology (Groves *et al.*, 2009), which can be expected to impinge on any forthcoming policy on SB as well.

Reflecting the Swiss activity in funding research in SB, several initiatives have been taken to sort out the role of SB with respect to its place in the regulatory landscape. In a meeting for biosafety advisory committees in Europe already organized in 2006 together with the Dutch COGEM (COGEM, 2011), the Swiss Expert Committee for Biosafety (SECB) considered SB to be similar to, or a subdiscipline of, gene technology. Hence, the criteria for the risk assessment of gene technology also apply to SB. Current work in SB was considered to take place mainly in contained facilities such as laboratories and bio-incubators, set up in accordance with the Swiss Federal Law on Gene Technology and the Ordinance on the Contained Use of Organisms that have proven safe in practice (sc | nat, 2006).

A somewhat different picture emerged from an expert report the ECNH published more recently. It defined SB as a relatively new field of research with an objective to turn biotechnology into an engineering discipline, paying special attention to the potential of SB to create new, artificial life forms (CH-ECNH, 2010). Accordingly, the current state of synthetic biology was dominated by visions, uncertainties, lack of knowledge and limited possibilities for risk analysis. Whether existing legal provisions for GMOs are sufficient for SB, therefore, would be difficult to establish (CH-ECNH, 2010). In their proposal for dealing with this situation the committee suggested generating knowledge step by step. Based on experiences with GMOs in the environment, situations implying potential risks should be considered carefully and repeatedly. They advocated separating risk description and analysis from risk evaluation, taking into account "duties of care" for risk management. This implies two responsibilities: (i) to anticipate possible consequences and damage potentials based on current knowledge and (ii) to consider necessary precautions to prevent damage. The committee also suggested distinguishing biosafety from biosecurity risks.

In Germany a widely acclaimed statement by a major science funging agency and two learned societies on synthetic biology said that biosafety and biosecurity risks were covered by current German laws (DFG-acatech-Leopoldina, 2009). Since the field is highly dynamic, however, they recommended to assign the Central Commission for Biological Safety (Zentrale Kommission für die Biologische Sicherheit; ZKBS) to carry out a scientific monitoring. The monitoring process should use the available expertise of the ZKBS to critically follow latest developments in SB. They also recommended to clearly define new methods for risk assessment for synthetic organisms that cannot be compared to existing natural reference organisms.

5.2.4 Outlook

There is a long-term challenge looming for policy makers and regulators. At a certain but yet unknown point in the not so distant future, the current set of regulatory and governance options might turn out to be insufficient, at least in part, to cover all SB techniques and possible fields of applications. In this case, the regulatory system could either be fine-tuned and tinkered with to include the latest developments, or undergo a major revision to take into account the dynamic and rapidly changing technological capabilities. Whether it is fine-tuning or a major overhaul, the key aspect in deploying any kind of change to the status quo of the regulatory practice is twofold: (i) to determine adequate objects of concern, assessment methods and criteria and (ii) to get the institutional setting right that should keep an eye on the developments.

Regarding objects, methods and criteria, steps have been made towards a proposal. For example, in their "priority paper" SYNBIOSAFE researchers[92] summarized possible future demands with respect to safety standards, in particular new methods in risk assessment to decide whether a new synthetic biology technique or application is safe enough (Schmidt *et al.*, 2009). They identified the following areas:

- DNA-based biocircuits consisting of a larger number of DNA "parts";
- The survivability and evolvability of novel minimal organisms used as platform or chassis biocircuits;
- "Exotic" biological systems based on an alternative biochemical structure (xenobiology).

In addition, synthetic biology itself may contribute to overcoming biosafety problems by facilitating the design of safer biosystems, for example, by:

- Designing less competitive organisms by changing metabolic pathways;
- Introducing metabolic pathways with a dependency on external biochemicals;
- Designing evolutionary robust biological circuits;
- Using alternative biological systems to avoid, for example, gene flow to and from wild species;
- Designing proto-cells that lack key features of living entities.

92) SYNBIOSAFE was the first European project on safety and ethical aspects of synthetic biology, see: www.synbiosafe.eu/

A third aspect addressed was skill diffusion (e.g., through do-it-yourself SB to amateurs or biohackers). Accordingly, everyone using the resources of SB should have awareness of and training in relevant biosafety techniques. In addition, appropriate mechanisms (laws, codes of conduct, voluntary measures, access restrictions to key materials, institutional embedding, mandatory reporting to Institutional Biosafety Committees) should be in place to avoid unintentional harm.

Regarding the institutional setting, the avenue is even less clear. It will not be easy to define, put together and establish a possibly permanent entity, whose responsibility would be to monitor and permanent assess SB developments and the immediate challenges they possibly could pose to the regulatory system. Such a monitoring entity could be established at the national level (like in Germany, where the Central Commission for Biological Safety was suggested to do this job), supranational level (i.e., the European Commission) or even at global level (such as the United Nations).

This would entail a rather substantial commitment from the side of the national organizations as well as a diligent construction of the committees since they need to retain both their independence and credibility. The implementation of the pertaining proposals to rely on monitoring, therefore, might be taken with a grain of salt. In addition, whether the various committees in charge would be able to agree on a common set of objects, methods and criteria remains a matter of speculation. Prior endeavors, for example, in harmonizing regulation on agricultural biotechnology, were not very promising. Nevertheless, if SB is going to fulfill its promises, early considerations on establishing practicable rules worldwide would be necessary.

5.3 China

5.3.1 Introduction

Synthetic biology (SB) is an emerging field in China with a handful of research groups currently active in SB-related projects, mainly funded by public funding and some limited funding from industry for biofuel research (Pei, Schmidt *et al.*, 2011a). SB has been considered as a priority area by the Ministry of Science and Technology (MOST) and listed as one of the 22 strategic technology issues that are keys to China's modernization reported in "Technological Revolution and China's Future–Innovation 2050" by the Chinese Academy of Science (CAS; CAS roadmap, 2007). It is highly expected that SB will promote the research in basic science, but more importantly will bring innovations in practical applications. Most of the current SB-related research activities in China serve these purposes. The focuses of the current research activities are mainly to develop useful applications using SB approaches and to enhance knowledge about biological systems. There are high expectations on synthetic biology derived applications for biofuels, bio-based chemicals, bioremediation and new types of medicine. Some of these research activities are listed in Table 5.7.

The Chinese laws and guidelines cover many subfields of biological research as well as those of medical research. The national regulations and institution-based review on scientific activities in general are applied to SB-related research. The current research governance model in China is based on scientifically informed, evidence-based approaches that are in general thought to be sufficient to cope with the current state of the art of the field. Many of these regulations were produced in recent years within the rapid development of the Chinese research community in these fields and most of them were drawn based upon the international guidance, such as the International Conference on the Harmonization of Technical Requirements for the Registration of Pharmaceuticals for Human Use, Good Clinical Practice Guidelines, the Cartagena Protocol on Biosafety for Living Modified Organisms (LMOs) and the World Health Organization handbook for biosafety. It is likely that the Chinese regulation on SB will also be developed based on an international guideline if existed. Most of the science policy is approved by the State Steering Committee of Science and Technology and Education. It is the committee that indirectly guides the work of three organizations (National Natural Science Foundation of China, aka NSFC, CAS, the Ministry of Education) and MOST. MOST is the most important national body to develop regulations in science and technology policy. The guidelines promulgated by MOST have a nationwide scope. The Chinese legislations relevant for SB are listed in Table 5.8.

5.3.2 General Provisions

So far, China has not yet developed SB-specific regulations. There are some existing regulations applied to SB. To obtain the second strategic objectives by year 2000 and to transfer China into one of the moderately developed countries in the middle of the twentyfirst century, a Circular of the State Council on Transmission of the State Medium- and Long-term Program on Science and Technology Development was announced on March 8, 1992.[93] Subsequently, the Law of the People's Republic of China on Science and Technology Progress, adopted at the Second Meeting of the Standing Committee of the Eighth National People's Congress on July 2, 1993, entered into force as of October 1, 1993.[94] It specified that the state should lie out guidance for the scientific and technological research and development and should establish a modernized scientific and technological research and development system, in accordance with the demands of economic construction and scientific and technological progress. Some of the ongoing SB projects have been assessed and funded by the programmes initiated by the law (Pei, Schmidt *et al.*, 2011a; Pei, Schmidt *et al.*, 2011b).

93) http://www.asianlii.org/cgi-bin/disp.pl/cn/legis/cen/laws/cotscototsmalposatd1146/cotscototsmalposatd1146.html?stem=0&synonyms=0&query=biology

94) http://www.asianlii.org/cn/legis/cen/laws/lotprocosatp607/

Table 5.7 Examples of some synthetic biology research in China.

	Research project	Institute
Genetic circuit/metabolic engineering		
Biofuel	To engineer cyanobacteria as a chassis organism to produce fatty acid-based biofuels	Qingdao Institute of Bioenergy and Bioprocess Technology, Chinese Academy of Science (CAS)
	Butanol	Key Laboratory of Synthetic Biology (KLSB), CAS
	Bioreactor for biomass energy and bio-based chemicals	State Key Laboratory of Bioreactor Engineering (SKL-BE), East China University of Science and Engineering
Bio-based chemicals	Polyhydroxyalkanoate	Tsinghua University
	Antibiotics (rifamycin, vancomycin)	KLSB, CAS
	Fine chemicals (riboflavin, succinate)	Tianjin University
	Artificial synthetic cell factory (Project 973)	Institute of Microbiology, CAS
Biosensors and bioremediation	Detoxification of pesticides by surface display of heterogeneous molecules on host microbes	SKL of Integrated Management of PIR, Institute of Zoology, CAS
	DNAzyme-based microarray to detect multiple metal ions	SKL-BE, East China University of Science and Engineering
	Engineering of *R. erythropolis* for bioremediation	SKL of Microbial Technology, Shandong University
Genetic control elements	Minimal model system in response to stimuli explored in a unifying framework	Nanjing University
	Unmarked genetic modification in *P. pastoris*	KLSB, CAS
	Reporter-guided mutant selection (RGMS)	Institute of Microbiology, CAS
Minimal genomes		
Bioinformatics approach	MyBASE for genome polymorphism and gene function studies of *Mycobacterium*	Institute of Psychology, CAS
	Database of essential genes (DEG) for survival of organisms	Tianjin University
Reduced genomes of model organisms	Engineered *P. putida* with reduced oxidation activities on fatty acids by deleting β-oxidation-related genes	Tsinghua University
	Minimal genome of *E. coli* and its synthesis (Project 973)	KLSB, CAS

Table 5.7 (*Continued*)

	Research project	Institute
Chemical synthetic biology		
Unnatural proteins	Monomer protein composed of β-α-β folds (two parallel β strands connected by an α helix)	Peking University
	β-α-β Structural motifs synthesized to build blocks in protein structures containing parallel β-sheets	Peking University
Proteins containing unnatual amino acids	*M. jannaschii* tRNA(Tyr)/tyrosyl-tRNA synthetase pair engineered to incorporate unnatural amino acids into proteins in *E. coli*	China Pharmaceutical University
Protocells		
Basic research	Use protocells to study energy conversion in cells	Peking University
Bioparts of protocell	DNA motor-driven nanopore switch for protocell	Peking University
DNA synthesis		
Improving DNA synthesis methods	PCR-based two-step synthesis for long DNA fragments	Shanghai Academy of Agricultural Sciences
	Isothermal DNA synthesis–isothermal unidirectional elongation method (IUEM)	Shanghai JiaoTong University
For basic research	Directed *in vitro* evolution of reporter genes based on semi-rational design and high-throughput screening	Shanghai Academy of Agricultural Sciences
	Codon-optimized recombinant xylanase gene	Shanghai Academy of Agricultural Sciences

Unlike the other well established research fields, such as genetically modified organisms GMOs and nanotechnology, major breakthroughs for the Chinese SB society are needed from frontline research to give the whole research field a boost. The majority of the Chinese scientific community knows very little about the progress of SB, and in fact that no dedicated regulation has been established to guide the research activities of SB. Currently SB-related research is regulated as other life science research – by national regulation and institution-based reviews of scientific activities in general. Although it is assumed that current scientifically informed evidence-based approaches to research governance are sufficient to cope with the current state of SB, an ongoing review on the current regulations to SB is needed based on overall progress.

Table 5.8 Government bodies involved the regulation of synthetic biology in China.

Government body	Abbreviation	Area of responsibility
State steering committee of science, technology and education		Most of the science policy is approved by the Committee and indirectly guide the work of MOST, NSFC, CAS and MOE
Ministry of science and technology	MOST	To develop national regulations in science and technology policy and as funding agency
		To provide guideline for recombinant DNA technology and biosecurity (reference list)
National natural science foundation of China	NSFC	To provide fund for basic research
Chinese academy of science	CAS	To run research and as potential funding agency
Ministry of education	MOE	To manage and regulate research in universities
Ministry of environmental protection	MOEP	To manage and regulate environmental effect of SB
Ministry of agriculture	MOA	To regulate on research on animal and plant
		To regulate on GMOs
		Responsible for biosafety of GMOs
		To regulate biosecurity (reference list)
Ministry of health	MOH	To regulate research related to human health; responsible for the biosafety on pathogenic organisms and biosecurity (reference list)
Ministry of foreign trade economic cooperation	MOFTEC	To regulate GMOs products

The research of SB is developed based on genetic engineering (GE) focusing on the whole system of genes and their products, thus there is no clear boundary between these two technologies. The GMOs have been developed successfully based on genetic engineering techniques and some of the GMOs are now in commercial use. The principles of these two research fields are somewhat different, but share many similarities (Zhang, 2009). It is argued that there is not much difference from SB to GE within the next 25 years regarding to its risk to human health and environment (Furger and Schweiz, 2007). One can expect that regulations on GE/GMOs may be used as a reference on the governance of SB at this moment with necessary revision. China is among the first countries commercializing GMO products; and biosafety regulations for GMOs have been established. We will review the legislation on GE/GMOs in China and their possible adaptations for risk management in SB.

MOST, formerly called the State Science and Technology Commission, is the major funding agency of biotechnology in China. It released the first biosafety regulation, "Safety Administration Regulation on Genetic Engineering", in 1993 (Song, 1993). It provided Chinese guidelines for using the recombinant DNA technology. It composed the articles on general safety principles, risk evaluation, safety control measures, and legal responsibilities. The National Genetic Engineering Biosafety Council was established which was responsible for the supervision and coordination of activities related to the safe and responsible for oversight the genetic engineering research. After this Regulation was issued, the Ministry of Agriculture (MOA) issued the "Implementation Regulation on Agricultural Genetic Engineering" in 1996, providing detailed regulation on GMOs for research and commercial production. In 1997, the National Agricultural GMOs Biosafety Committee was established, acting as the expert committee for MOA on biosafety regulations. In May 2001, China's State Council released a new guideline "Regulation on Safety Administration of Agricultural GMOs" (The State Council, 2001). And in early 2002, MOA issued three detailed regulations on the biosafety management, trading and labeling of agricultural products derived from GMOs (Huang and Wang, 2002). This regulation included several important changes to the existing procedures and listed regulatory responsibilities of post-commercialization in detail. Such changes included new processing regulations for GM products, labeling requirements for marketing and local- and provincial-level GMO monitoring guidelines. In the meantime, the Ministry of Health (MOH) also promulgated its first regulation on GM food safety in April 2002, and this came into effect in July 2002. The MOA is the primary department in charge of the formulation and implementation of biosafety regulations on agricultural GMOs and their commercialization. In order to coordinate the regulations from different ministries, the State Council established a Ministerial Meeting comprised by the representatives from MOA, MOST, MOH, the State Development Planning Commission (SDPC), the Inspection and Quarantine Agency, the Ministry of Environmental Protection (MEP) and the Ministry of Foreign Trade Economic Cooperation (MOFTEC). Currently, the development of biotechnology focuses on GMOs for agricultural purposes in China, such as transgenic crops. Implementation of the regulation is operated by the Biosafety Office of Agricultural GMOs under MOA. The National GMOs Biosafety Committee plays an important role in the biosafety management of agricultural GMOs and has two meetings each year to evaluate biosafety assessment applications related to laboratory research,field trials, environmental release and the commercialization of GMOs for agriculture. It provides recommendations to the Biosafety Office of Agricultural GMOs of MOA based on the results of biosafety assessments. And this office may make the final decisions based on these recommendations. The MOH is responsible for food safety management of biotechnology products, especially from GMOs. The Appraisal Committee, consisting of experts nominated from the fields of food health, nutrition and toxicology by MOH, is responsible for reviewing and assessing GM foods. In 2005, provincial biosafety management offices affiliated to provincial agricultural bureaus were established in all Chinese provinces. These biosafety management offices are responsible for local statistic data collection and monitoring the ongoing research and commercialization in the agricultural biotechnology industry. These local offices are in charge of evaluation and ruling on all applications of GM related research, field trials and commercialization. Only applications approved by provincial biosafety management offices can be submitted to the National Biosafety Committee for further assessment.

A Law of the People's Republic of China on the Prevention and Treatment of Infectious Diseases was issued on August 28, 2004, to prevent and control the outbreak and spread of infectious diseases, to ensure public health. The outbreak of SARS in the year 2003 highlighted the urgent need for infectious disease control in China. A policy was implemented with emphasis on prevention, combining prevention with treatment, exercising classified control and relying on science and the masses.[95] The infectious agents are classified into Classes A, B and C. The reference list of animal and human pathogenic microorganisms includes 123 species and 380 species, respectively. The health administration department under the State Council (the chief administration authority in China) is in charge of the work of preventing and treating infectious diseases and supervising related activities and disease control nationwide.[96] Besides this regulation, the State Council issued a dedicated guideline for laboratory biosafety – "General Biosafety Standard for Microbiology and Biomedical Laboratories" (W233-2002) in 2002[97] and "Veterinary Laboratory Biosafety Guidelines" in 2003.[98] For SB-related research involving pathogenic microbes, including the microorganism itself or the related medical applications, the regulation of such research activities will follow in the regulation provided by those guidelines.

Creating SB engineered microorganisms for bioremediation is one of the promising SB approaches that are currently under extensive investigations both in China and around the world. Thus, environmental release of SB-derived living organisms will pose a great challenge for proper regulation. The Environmental Protection Law of the People's Republic of China was issued in December 1989, pointing out that the regulations for environmental protection issued by the State must be incorporated into national economic and social development plans; the state should adopt economic and technological policies and measures which are favorable for environmental protection so as to coordinate the work of environ-

95) http://www.asianlii.org/cgi-bin/disp.pl/cn/legis/cen/laws/patoidl487/patoidl487.html?stem=0&synonyms=0&query=pathogenic

96) http://news.xinhuanet.com/zhengfu/2004-11/29/content_2271255.htm

97) http://biosafety.sysu.edu.cn/administer/national/200804/79.html

98) http://xn.sdmyxy.cn:8013/Article/ShowArticle.asp?ArticleID=20

mental protection with economic construction and social development.[99] The MEP is in charge of establishing the national standards for environmental quality, where local standards for the environmental quality of items not specified in the national standards for environment quality shall be established regionally and reported to the competent department of MEP for the record. "Methods for the Biosafety Environmental Management of Pathogenic Microbiology Laboratories" was issued by the MEP in 2006.[100] This law specified that biosafety laboratories would be classified in four levels (BSL-1, 2, 3, 4); within this classification, research involving highly pathogenic microorganisms can only be conducted in certified BSL-3 and BSL-4 laboratories. It is likely that SB research involving environmental release will need to follow both national and regional regulations.

5.3.3 Biosecurity and Dual Use

Biosecurity of SB mainly deals with issues of dual use of the techniques, particularly the possibility of SB to create highly pathogenic living organisms that can be used as bioweapons (Fink Committee, 2004; National Science Advisory Board for Biosecurity (NSABB) 2008). Among the Chinese research community, dual use is not just for SB techniques, but is common for any other science and technology (Pei, Schmidt *et al.*, 2011a). Concerns about research on highly pathogenic organisms are not issues for the Chinese scientists. That is due to the fact that all research on highly pathogenic organisms is under stringent regulation in China. Only very few research institutes are authorized to work on them. Back in the 1980s, regulations were issued on the proper storage of selected strains, for instance, "Methods on the Trial Management of the Preservation of Veterinary Microbial Strains" by MOA in 1980 (replaced by updated regulations in 2008),[101] "Methods of Management of the Preservation of Medical-Microbiology Strains in China" by MOH in 1985[102] and "Rules on the Management of the Preservation of Microbial Strains in China" by MOST in 1986.[103] Under these regulations, only limited culture collection centers appointed by MOA and MOH can work on agents on the reference list. These dedicated centers are equipped with physical security and certified facilities to handle the selected agents. Other institutes with historical collections were no longer authorized to maintain their collections on the reference list. To strengthen the export control of dual use biological agents and related equipment and technologies and to safeguard the State security and social and public interests, the Regulations of the People's Republic of China on the Export Control of Dual-Use Biological Agents and Related Equipment and Technologies were promulgated and came into force in December 2002.[104] The

99) http://www.china.org.cn/english/environment/34356.htm
100) http://www.sepa.gov.cn/info/gw/juling/200603/t20060308_74730.htm
101) http://www.cvcc.org.cn/102578/80006.html
102) http://www.pxwsj.com/wswjz.htm
103) http://www.lawxp.com/statute/s980874.html
104) http://www.asianlii.org/cgi-bin/disp.pl/cn/legis/cen/laws/rotprocoecod-baareat1147/rotprocoecodbaareat1147.html?stem=0&synonyms=0&query=chinese%20biosafety)

export of dual use biological agents and related equipment and technologies should be in accordance with the relevant laws, the administrative regulations of the State and these Regulations, and they should not imperil State security and social and public interests.

5.3.4
Options for Adapting and Improving Regulations

As an emerging research field in China, it is not difficult to realize there will be many new challenges brought by SB, particularly issues of biosafety and biosecurity. Existing research on the biosafety of SB is scarce in China, which is lagging behind application-oriented research. There is therefore hardly any technical support for safety evaluation. The growing public concerns on GMO products in China calls for better regulations on SB to reassure the public about the safety of the future products derived from SB techniques. All these indicate the urgent need to build a proper regulation framework to cope with the development of this field. Thus, we propose here some recommendations for specific regulations on SB:

1) **To promote multi-disciplinary research.** Due to the multi-disciplinary nature of SB, it is necessary to develop an interdisciplinary research center, to promote the basic research of SB in China. Although a few dedicated centers for SB were set up recently, they were more focused on serving the purposes to produce biological products of interest (such as biomedicines, biofuels and biobased products), lacking integration with research teams from other disciplines. In addition, a multi-disciplinary center of SB can provide multi-disciplinary training and promote intellectual knowledge exchange. This will in turn promote the development of SB among the Chinese scientific community.
2) **To establish norms on safety assessment and risk analysis.** According to practical experience of the safety management of GMOs, the risk assessment on SB products should be carried out based on their classification according to risk degree and familiar principle. The key links of safety management and critical points of risk should be investigated.
3) **To develop practical guidelines to improve their implementability.** China has drawn up some safety assessments and guidelines which play an important role in the safety management of biological and medical research. Based on the experience of executing the existing GMO regulatory framework, practical guidance in detail for SB research is needed for each tier of oversight (from grant application and continuous supervision to on-site inspection) to improve implementation of the regulations.
4) **To set up a national committee to monitor SB-derived products.** A national committee should be set up to monitor the synthetic products from R&D, through production, to on the market, to ensure the safety measurements of SB are in place. Citing the experience of other biosafety committees set up

for agriculture and for GMOs, a national SB biosafety committee should be also set up, composed by experts from different fields (such as public health, environmental science, food science and industry).

5) **To improve the accountability of management within institutions.** Research and development institutions are the primary sites to implement the safety regulations. To improve the accountability of these institutes is critical to ensure the implementation of SB safety measures and to facilitate SB research (The State Council, 2011; The ministry of agriculture, 2002). All research institutes should be responsible for providing proper training on biosafety measurements to all the personnel involved (from researchers to supporting personnel). Internal supervisors should be appointed to make sure the safety rules are followed properly.

6) **To build a framework to coordinate regulations among administrative bodies.** The safety issues of SB involve regulations from many administrative bodies, such as those from MOST, MOA, MOH and MEP. Cooperation and communication among them is critical to build consistent regulations on SB. It is necessary to develop a coordinated approach by establishing a framework to clarify the responsibility of every administration agency, to avoid any redundancy on over-administration.

7) **To enhance the scientific outreach of SB, focusing on biosafety education.** The public perception of SB is important not only for SB, but also for its safety management. Therefore, it is necessary to enhance the scientific outreach of SB, for instance, setting up public dialog to discuss issues around SB, from basic sciences to ethical and social issues. Special websites on SB should be built to introduce this emerging science to the public and interested groups of its research and development, biosafety and relevant management. Moreover, scientific outreach needs to be strengthened so that communication between the government and the scientific community, the media and the general public can be initiated. This will help to ensure the healthy development of SB.

5.3.5 Outlook

Synthetic biology, although still in an early developmental stage, has already been established in China, as shown by current research activities. Not only is more dedicated funding needed to support this promising research, but also dedicated research regulations are needed to foster the future development of SB in China. The current regulations are for institutionally based research with guidelines issued by MOST, MOH, MOA and MEP. The research is regulated depending where the research activities are conducted and where the funding comes from. These guidelines of research governance have covered most of the ongoing research activities and are able to cope with the current state of the art of SB in the field, which is in its early stage of development. Yet active monitoring the progress of the field is needed for an up to date, transparent and evolving regulation of an emerging field.

The biosafety issues are important to ensure the successful future of an emerging field. Lessons learned from the commercialization of GMOs products have set good examples for possible innovations from other sectors. China has made great progress in improving its biosafety standards – practical laboratory biosafety procedures, standards for containment and guideline for facilities. However, the implementation of biosafety rules varies depending on the setting in which research occurs. The responsibility and self-regulation of host institutes need to be improved. Therefore, enhancing the accountability of the existing regulations will be crucial to ensure the biosafety of SB.

It is known that the risk analysis of SB is a challenge – these are risks of great uncertainty. A perceived gap between the theoretical and physical foundation for living organisms makes it difficult for risk assessment. Thus, regulatory processes should be updated regularly as the field progresses. Research focusing on risk assessment should gain more support. With more expenditure on research in the new five-year plan, Chinese research activities on SB will definitely increase. It is important that proper regulations are in place to cope with developments in the field. In addition, promoting the scientific outreaching and public dialog on SB will help to ensure SB advances in a way that ethical, legal and social implications are all taken into consideration.

References

5.1. United States of America

See footnotes in text for links to United States regulations and laws.

Street, P. (2007) Constructing risks: GMOs, biosafety and environmental decision-making, in *The Regulatory Challenge of Biotechnology* (ed. H. Somsen), Edward Elgar Publishing, Cheltenham, p. 186.

5.2. EUROPE

Bedau, M.A., Parke, E.C., Tangen, U., and Hantsche-Tangen, B. (2009) Social and ethical checkpoints for bottom-up synthetic biology, or protocells. *Syst. Synth. Biol.*, **3**, 65–75.

Benner, S.A., and Sismour, A.M. (2005) Synthetic biology. *Nat. Rev. Genet.*, **6**, 533–543.

BTWC (1975) Convention on the Prohibition of the Development, Production and Stockpiling of Bacteriological (Biological) and Toxin Weapons and on Their Destruction.

Budisa, N. (2004) Prolegomena to future experimental efforts on genetic code engineering by expanding its amino acid repertoire. *Angew. Chem. Int. Ed.*, **43**, 6426–6463.

Bugl, H., Danner, J.P., Molinari, R.J., Mulligan, J.T., Park, H.-O., Reichert, B., Roth, D.A., Wagner, R., Budowle, B., Scripp, R.M., Smith, J.A.L., Steele, S.J., Church, G., and Endy, D. (2007) DNA synthesis and biological security. *Nat. Biotechnol.*, **25**, 627–629.

Bunn, S. (2008) Synthetic biology Postnote 298, 1–4.

Campos, L. (2009) That was the synthetic biology that was, in *Synthetic Biology: The Technoscience and Its Societal Consequences* (eds M. Schmidt, A. Kelle, A. Ganguli-Mitra, and H. deVriend), Springer, p. 186.

Canton, B., Labno, A., and Endy, D. (2008) Refinement and standardization of synthetic biological parts and devices. *Nat. Biotechnol.*, **26**, 787–793.

Carlson, R. (2003) The pace and proliferation of biological technologies. *Biosecur. Bioterror.*, **1**, 203–214.

Carlson, R. (2009) The changing economics of DNA synthesis. *Nat. Biotechnol.*, **27**, 1091–1094.

Carlson, R.H. (2010) *Biology Is Technology: The Promise, Peril, and New Business of Engineering Life*, Harvard University Press, Cambridge, MA.

Carlson, S.R., Rudgers, G.W., Zieler, H., Mach, J.M., Luo, S., Grunden, E., Krol, C., Copenhaver, G.P., and Preuss, D. (2007) Meiotic transmission of an *in vitro*-assembled autonomous maize minichromosome. *PLoS Genet.*, **3**, 1965–1974.

Carr, P.A., and Church, G.M. (2009) Genome engineering. *Nat. Biotechnol.*, **27**, 1151–1162.

Cello, J., Paul, A.V., and Wimmer, E. (2002) Chemical synthesis of poliovirus cDNA: generation of infectious virus in the absence of natural template. *Science*, **297**, 1016–1018.

CH-ECNH (2010) *Synthetic Biology – Ethical Considerations* (ed. ECNH FECoN-HB), CH-ECNH.

CH-ECNH (2011) Federal Ethics Committee on Non-Human Biotechnology.

CH-Epidemiengesetz (1970) *Bundesgesetz über die Bekämpfung übertragbarer Krankheiten des Menschen (Epidemiengesetz)* (ed. Eidgenossenschaft BdS), CH-ECNH.

CH-FOEN (2011) *Environmental Law: Gene Technology* (ed. FOEN FOftE), CH-ECNH.

Ch-FOPH (2011) Federal Office of Public Health.

CH-FrSV (2008) *Verordnung über den Umgang mit Organismen in der Umwelt (Freisetzungsverordnung, FrSV)* (ed. Eidgenossenschaft BdS), CH-ECNH.

CH-GTG (2003) *Bundesgesetz über die Gentechnik im Ausserhumanbereich (Gentechnikgesetz, GTG)* (ed. Eidgenossenschaft BdS), CH-ECNH.

CH-SAMV (1999) Verordnung über den Schutz der Arbeitnehmerinnen und Arbeitnehmer vor Gefährdung durch Mikroorganismen (SAMV).

CH-SECB (2011) *Regulatory Documents, Guidelines and Additional Information* (ed. Biosafety SECf), CH-ECNH.

CH-Sicherheit (1996) *Verordnung über die Eidgenössische Fachkommission für biologische Sicherheit* (ed. Eidgenossenschaft BdS), CH-ECNH.

CH-USG (1983) *BUNDESGESETZ ueber den Umweltschutz (Umweltschutzgesetz, USG)* (ed. Eidgenossenschaft BdS), CH-ECNH.

CH-UVG (1981) *BUNDESGESETZ ueber die Unfallversicherung (UVG)* (ed. Eidgenossenschaft BdS), CH-ECNH.

COGEM (2011) EVALUATION COGEM 2011. The Netherlands Commission on Genetic Modification.

Danchin, A. (2009) Information of the chassis and information of the program in synthetic cells. *Syst. Synth. Biol.*, **3**, 125–134.

Declercq, R., Van Aerschot, A., Read, R.J., Herdewijn, P., and Van Meervelt, L. (2002) Crystal structure of double helical hexitol nucleic acids. *J. Am. Chem. Soc.*, **124**, 928–933.

de Lorenzo, V. (2010) Environmental biosafety in the age of synthetic biology: do we really need a radical new approach? Environmental fates of microorganisms bearing synthetic genomes could be predicted from previous data on traditionally engineered bacteria for *in situ* bioremediation. *BioEssays*, **32**, 926–931.

Deplazes, A. (2009) Piecing together a puzzle. An exposition of synthetic biology. *EMBO Rep.*, **10**, 428–432.

Deplazes, A., and Huppenbauer, M. (2009) Synthetic organisms and living machines: positioning the products of synthetic biology at the borderline between living and non-living matter. *Syst. Synth. Biol.*, **3**, 55–63.

DFG-acatech-Leopoldina (2009) *Synthetische Biologe: Stellungnahme*, DFG, Berlin, Germany, p. 21.

Dymond, J.S., Richardson, S.M., Coombes, C.E., Babatz, T., Muller, H., Annaluru, N., Blake, W.J., Schwerzmann, J.W., Dai, J., Lindstrom, D.L., Boeke, A.C., Gottschling, D.E., Chandrasegaran, S., Bader, J.S., and Boeke, J.D. (2011) Synthetic chromosome arms function in yeast and generate

phenotypic diversity by design. *Nature*, **477**, 471–476.
EASAC (2010) *Realising European Potential in Synthetic Biology: Scientific Opportunities and Good Governance* (ed. Council EASA), German Academy of Sciences Leopoldina.
EASAC (2011) *Synthetic Biology: An Introduction* (ed. Council EASA), EASAC.
Elowitz, M.B., and Leibler, S. (2000) A synthetic oscillatory network of transcriptional regulators. *Nature*, **403**, 335–338.
Endy, D. (2005) Foundations for engineering biology. *Nature*, **438**, 449–453.
European Commission (1976) Council Directive 76/768/EEC.
European Commission (1982) Council Directive 82/894/EEC.
European Commission (1990a) Council Directive 90/219/EEC.
European Commission (1990b) Council Directive 90/385/EEC Official Journal of the European Communities.
European Commission (1993) Council Directive 93/42/EEC. Official Journal of the European Communities, pp. 1–43.
European Commission (1995) Directive 95/46/EC.
European Commission (1998) Council Directive 98/81/EC. Official Journal of the European Communities, pp. 13–31.
European Commission (2000a) Council Directive 2000/29/EC.
European Commission (2000b) Council Directive 2000/54/EC. Official Journal of the European Communities, pp. 21–43.
European Commission (2000c) Council Regulation (EC) No 1334/2000.
European Commission (2001a) Council Directive 2001/18/EC. Official Journal of the European Communities, pp. 1–38.
European Commission (2001b) Council Directive 2001/83/EC. Official Journal of the European Communities, pp. 62–128.
European Commission (2002a) Directive 2002/58/EC.
European Commission (2002b) Directive 2002/98/EC.
European Commission (2003a) COMMISSION DIRECTIVE 2003/63/EC.
European Commission (2003b) COMMISSION DIRECTIVE 2003/94/EC.
European Commission (2003c) REGULATION (EC) No 1946/2003.
European Commission (2004a) Directive 2004/23/EC.
European Commission (2004b) REGULATION (EC) No 726/2004.
European Commission (2005) The European Union Counter-Terrorism Strategy.
European Commission (2006a) Directive 2006/121/EC.
European Commission (2006b) Regulation (EC) No 1907/2006.
European Commission (2007a) Regulation (EC) No 1394/2007.
European Commission (2007b) SYNTHETIC BIOLOGY A NEST PATHFINDER INITIATIVE.
European Commission (2009a) COUNCIL REGULATION (EC) No 428/2009. Official Journal of the European Union.
European Commission (2009b) Ethically speaking.
European Commission (2009c) Ethics of synthetic biology. Publications Office of the European Union, Luxembourg.
European Commission (2009d) The European Group on Ethics of science and new technologies (EGE) Opinion on the ethics of synthetic biology.
European Commission (2010) Synthetic Biology From Science to Governance. A workshop organised by the European Commission's Directorate-General for Health & Consumers, Brussels.
Furger, F., and Schweiz, F. (2007) From Genetically Modified Organisms to Synthetic Biology: Legislation in the European Union, in Six Member Countries and in Switzerland. Working Papers for Synthetic Genomics: Risks and Benefits for Science and Society.
Garfinkel, M.S., Endy, D., Epstein, G.L., and Friedman, R.M. (2007) Synthetic Genomics | Options for Governance. The J. Craig Venter Institute.
Gibson, D.G., Benders, G.A., Andrews-Pfannkoch, C., Denisova, E.A., Baden-Tillson, H., Zaveri, J., Stockwell, T.B., Brownley, A., Thomas, D.W., Algire, M.A., Merryman, C., Young, L., Noskov, V.N., Glass, J.I., Venter, J.C., Hutchison, C.A., 3rd, and Smith, H.O. (2008) Complete chemical synthesis, assembly, and cloning of a Mycoplasma genitalium genome. *Science*, **319**, 1215–1220.
Gibson, D.G., Glass, J.I., Lartigue, C., Noskov, V.N., Chuang, R.Y., Algire, M.A.,

Benders, G.A., Montague, M.G., Ma, L., Moodie, M.M., Merryman, C., Vashee, S., Krishnakumar, R., Assad-Garcia, N., Andrews-Pfannkoch, C., Denisova, E.A., Young, L., Qi, Z.Q., Segall-Shapiro, T.H., Calvey, C.H., Parmar, P.P., Hutchison, C.A., Smith, H.O., and Venter, J.C. (2010) Creation of a bacterial cell controlled by a chemically synthesized genome. *Science*, **329**, 52–56.

Greber, D., and Fussenegger, M. (2007) Mammalian synthetic biology: engineering of sophisticated gene networks. *J. Biotechnol.*, **130**, 329–345.

Groves, C., Frater, L., Lee, R., Jenkins, H., and Yakovleva, N. (2009) *An Examination of the Nature and Application among the Nanotechnologies Industries of Corporate Social Responsibility in the Context of Safeguarding the Environment and Human Health*, University–Cardiff.

Hanczyc, M.M., and Szostak, J.W. (2004) Replicating vesicles as models of primitive cell growth and division. *Curr. Opin. Chem. Biol.*, **8**, 660–664.

Hartman, M.C., Josephson, K., Lin, C.W., and Szostak, J.W. (2007) An expanded set of amino acid analogs for the ribosomal translation of unnatural peptides. *PLoS ONE*, **2**, e972.

Henry, A.A., and Romesberg, F.E. (2003) Beyond A, C, G and T: augmenting nature's alphabet. *Curr. Opin. Chem. Biol.*, **7**, 727–733.

Herdewijn, P., and Marliere, P. (2009) Toward safe genetically modified organisms through the chemical diversification of nucleic acids. *Chem. Biodivers.*, **6**, 791–808.

HSE (2007a) HSE Horizon Scanning Intelligence Group Short Report Synthetic biology.

HSE (2007b) The SACGM Compendium of guidance Part 1. Health and Safety Executive.

HSE (2007c) The SACGM Compendium of guidance Part 2. Health and Safety Executive.

Hutchison, C.A., Peterson, S.N., Gill, S.R., Cline, R.T., White, O., Fraser, C.M., Smith, H.O., and Venter, J.C. (1999) Global transposon mutagenesis and a minimal Mycoplasma genome. *Science*, **286**, 2165–2169.

Ichida, J.K., Horhota, A., Zou, K., McLaughlin, L.W., and Szostak, J.W. (2005) High fidelity TNA synthesis by Therminator polymerase. *Nucleic Acids Res.*, **33**, 5219–5225.

IRGC (2010) Guidelines for the Appropriate Risk Governance of Synthetic Biology. International Risk Governance Council.

Itaya, M. (1995) An estimation of minimal genome size required for life. *FEBS Lett.*, **362**, 257–260.

Kelle, A. (2009) Ensuring the security of synthetic biology-towards a 5P governance strategy. *Syst. Synth. Biol.*, **3**, 85–90.

Koonin, E.V. (2000) How many genes can make a cell: the minimal-gene-set concept. *Annu. Rev. Genomics Hum. Genet.*, **1**, 99–116.

Lartigue, C., Glass, J.I., Alperovich, N., Pieper, R., Parmar, P.P., Hutchison, C.A., 3rd, Smith, H.O., and Venter, J.C. (2007) Genome transplantation in bacteria: changing one species to another. *Science*, **317**, 632–638.

Leduc, S. (1910) Théorie physico-chimique de la vie et générations spontanées.

Leduc, S. (1912) La biologie synthétique.

Legislation UK (1990) Environmental Protection Act 1990.

Legislation UK (1996) The Genetically Modified Organisms (Risk Assessment) (Records and Exemptions) Regulations 1996.

Legislation UK (1997) The Genetically Modified Organisms (Deliberate Release and Risk Assessment-Amendment) Regulations 1997.

Levskaya, A., Chevalier, A.A., Tabor, J.J., Simpson, Z.B., Lavery, L.A., Levy, M., Davidson, E.A., Scouras, A., Ellington, A.D., Marcotte, E.M., and Voigt, C.A. (2005) Synthetic biology: engineering *Escherichia coli* to see light. *Nature*, **438**, 441–442.

Loakes, D., and Holliger, P. (2009) Polymerase engineering: towards the encoded synthesis of unnatural biopolymers. *Chem. Commun. (Camb)*, 4619–4631.

Lu, T.K., Khalil, A.S., and Collins, J.J. (2009) Next-generation synthetic gene networks. *Nat. Biotechnol.*, **27**, 1139–1150.

Luisi, P.L. (2007) Chemical aspects of synthetic biology. *Chem. Biodivers.*, **4**, 603–621.

Luisi, P.L., Chiarabelli, C., and Stano, P. (2006) From never born proteins to minimal living cells: two projects in synthetic biology. *Orig. Life Evol. Biosph.*, **36**, 605–616.

Macnab, S., and Whitehouse, A. (2009) Progress and prospects: human artificial chromosomes. *Gene Ther.*, **16**, 1180–1188.

Mansy, S.S., and Szostak, J.W. (2008) Thermostability of model protocell membranes. *Proc. Natl. Acad. Sci. U. S. A.*, **105**, 13351–13355.

Mansy, S.S., Schrum, J.P., Krishnamurthy, M., Tobe, S., Treco, D.A., and Szostak, J.W. (2008) Template-directed synthesis of a genetic polymer in a model protocell. *Nature*, **454**, 122–125.

Marliere, P. (2009) The farther, the safer: a manifesto for securely navigating synthetic species away from the old living world. *Syst. Synth. Biol.*, **3**, 77–84.

Marliere, P., Patrouix, J., Doring, V., Herdewijn, P., Tricot, S., Cruveiller, S., Bouzon, M., and Mutzel, R. (2011) Chemical evolution of a bacterium's genome. *Angew. Chem. Int. Ed.*, **50**, 7109–7114.

May, M. (2009) Engineering a new business. *Nat. Biotechnol.*, **27**, 1112–1120.

Murtas, G. (2010) Internal lipid synthesis and vesicle growth as a step toward self-reproduction of the minimal cell. *Syst. Synth. Biol.*, **4**, 85–93.

Mushegian, A. (1999) The minimal genome concept. *Curr. Opin. Genet. Dev.*, **9**, 709–714.

National Institutes of Health (2010) *Proposed Actions under the NIH Guidelines for Research Involving Recombinant DNA Molecules (NIH Guidelines)*, vol. 75 (ed. SERVICES DOHAH), Federal Register, (77) p. 3.

Neumann, H., Wang, K., Davis, L., Garcia-Alai, M., and Chin, J.W. (2010) Encoding multiple unnatural amino acids via evolution of a quadruplet-decoding ribosome. *Nature*, **464**, 441–444.

Nielsen, P.E., and Egholm, M. (1999) An introduction to peptide nucleic acid. *Curr. Issues Mol. Biol.*, **1**, 89–104.

O'Malley, M.A., Powell, A., Davies, J.F., and Calvert, J. (2008) Knowledge-making distinctions in synthetic biology. *BioEssays*, **30**, 57–65.

Posfai, G., Plunkett, G., 3rd, Feher, T., Frisch, D., Keil, G.M., Umenhoffer, K., Kolisnychenko, V., Stahl, B., Sharma, S.S., de Arruda, M., Burland, V., Harcum, S.W., and Blattner, F.R. (2006) Emergent properties of reduced-genome *Escherichia coli*. *Science*, **312**, 1044–1046.

Presidential Commission for the Study of Bioethical Issues (2010) New Directions: The Ethics of Synthetic Biology and Emerging Technologies. Presidential Commission for the Study of Bioethical Issues, p. 192.

Rasmussen, S. (2009) *Protocells: Bridging Nonliving and Living Matter*, MIT Press, Cambridge, MA.

Rasmussen, S., Chen, L., Deamer, D., Krakauer, D.C., Packard, N.H., Stadler, P.F., and Bedau, M.A. (2004) Evolution. Transitions from nonliving to living matter. *Science*, **303**, 963–965.

Ro, D.-K., Paradise, E.M., Ouellet, M., Fisher, K.J., Newman, K.L., Ndungu, J.M., Ho, K.A., Eachus, R.A., Ham, T.S., Kirby, J., Chang, M.C.Y., Withers, S.T., Shiba, Y., Sarpong, R., and Keasling, J.D. (2006) Production of the antimalarial drug precursor artemisinic acid in engineered yeast. *Nature*, **440**, 940–943.

Royal Academy of Engineering (2009) Synthetic Biology: scope, applications and implications.

Royal Society (2007) Synthetic biology: Call for views.

sc | nat (2006) Synthetic Biology.

Schmidt, M. (2009) *Do I Understand What I Can Create? Biosafety Issues in Synthetic Biology. The Technoscience and Its Societal Consequences*, Springer Academic Publishing.

Schmidt, M. (2010) Xenobiology: a new form of life as the ultimate biosafety tool. *BioEssays*, **32**, 322–331.

Schmidt, M., and Pei, L. (2011) Synthetic toxicology: where engineering meets biology and toxicology. *Toxicol. Sci.*, **120** (Suppl. 1), S204–S224.

Schmidt, M., Ganguli-Mitra, A., Torgersen, H., Kelle, A., Deplazes, A., and Biller-Andorno, N. (2009) A priority paper for the societal and ethical aspects of synthetic biology. *Syst. Synth. Biol.*, **3**, 3–7.

Schoning, K., Scholz, P., Guntha, S., Wu, X., Krishnamurthy, R., and Eschenmoser,

A. (2000) Chemical etiology of nucleic acid structure: the alpha-threof uranosyl-(3′–>2′) oligonucleotide system. *Science*, **290**, 1347–1351.

SECO (2004) *Federal Act on the Control of Dual-Use Goods and of Specific Military Goods* (ed. Affairs SSSoE), SSSoE, Bern, 946202.

SECO (2005) *Ordinance on the Export, Import and Transit of Dual Use Goods and Specific Military Goods* (ed. Affairs SSSoE), SSSoE, Bern, 9462021.

Smolke, C.D. (2009) Building outside of the box: iGEM and the BioBricks Foundation. *Nat. Biotechnol.*, **27**, 1099–1102.

Stricker, J., Cookson, S., Bennett, M.R., Mather, W.H., Tsimring, L.S., and Hasty, J. (2008) A fast, robust and tunable synthetic gene oscillator. *Nature*, **456**, 516–519.

Szostak, J.W., Bartel, D.P., and Luisi, P.L. (2001) Synthesizing life. *Nature*, **409**, 387–390.

The Academy of Medical Sciences and Engineering (2007) Systems Biology: A Vision for Engineering and Medicine.

The Royal Society (2008) *Synthetic Biology: Scientific Discussion Meeting Summary*, The Royal Society.

Tigges, M., Denervaud, N., Greber, D., Stelling, J., and Fussenegger, M. (2010) A synthetic low-frequency mammalian oscillator. *Nucleic Acids Res.*, **38**, 2702–2711.

Tumpey, T.M., Basler, C.F., Aguilar, P.V., Zeng, H., Solorzano, A., Swayne, D.E., Cox, N.J., Katz, J.M., Taubenberger, J.K., Palese, P., and Garcia-Sastre, A. (2005) Characterization of the reconstructed 1918 Spanish influenza pandemic virus. *Science*, **310**, 77–80.

Umweltbundesamt (2004) 3.9. The Application of Genetically Modified Organisms (GMOs). State of the Environment 2004 7th Report on the State of the Environment in Austria. gugler print & media GmbH, Melk an der Donau.

Umweltbundesamt (2008) Übersicht über die österreichischen Rechtsgrundlagen für gentechnisch veränderte Organismen (GVO) und GVO-Produkte.

Walde, P. (2010) Building artificial cells and protocell models: experimental approaches with lipid vesicles. *BioEssays*, **32**, 296–303.

WHO (2004) *Laboratory Biosafety Manual*, 3rd edn, Whole Health Organization, Geneva.

Wimmer, E., Mueller, S., Tumpey, T.M., and Taubenberger, J.K. (2009) Synthetic viruses: a new opportunity to understand and prevent viral disease. *Nat. Biotechnol.*, **27**, 1163–1172.

Yang, Z., Hutter, D., Sheng, P., Sismour, A.M., and Benner, S.A. (2006) Artificially expanded genetic information system: a new base pair with an alternative hydrogen bonding pattern. *Nucleic Acids Res.*, **34**, 6095–6101.

Yang, Z., Sismour, A.M., Sheng, P., Puskar, N.L., and Benner, S.A. (2007) Enzymatic incorporation of a third nucleobase pair. *Nucleic Acids Res.*, **35**, 4238–4249.

5.3. CHINA

CAS roadmap (2007) Innovation 2050: Technology Revolution and the Future of China.

Fink Committee (2004) Biotechnology Research In An Age Of Terrorism: Confronting The Dual Use Dilemma.

Furger, F., and Schweiz, F. (2007) From Genetically Modified Organisms to Synthetic Biology: Legislation in the European Union, in Six Member Countries and in Switzerland. Working Papers for Synthetic Genomics: Risks and Benefits for Science and Society.

Huang, J., and Wang, Q. (2002) Agricultural biotechnology development and policy in China. *AgBio- Forum*, **5** (4), 122–135.

National Science Advisory Board for Biosecurity (NSABB) (2008) Strategic Plan for Outreach and Education on Dual Use Research Issues.

Pei, L., Schmidt, M., *et al.* (2011) New life in the laboratory. *The Biochemist*, **2011**, 14–18.

Pei, L., Schmidt, M., *et al.* (2011) Synthetic biology: An emerging research field in China. *Biotechnol. Adv.*, **29** (6), 804–814.

Song, J. (1993) Safety administration regulation on genetic engineering. The State Science and Technology Commission of the People's Republic of China.

The Ministry of Agriculture (2002-01-05) Agricultural genetically modified organisms safety evaluation management approach, http://www.gov.cn/gongbao/content/2002/content_61847.htm (accessed 1 November 2011).

The State Council (2001-05-23) Agricultural genetically modified organisms safety control regulations, http://www.gov.cn/gongbao/content/2001/content_60893.htm (accessed 1 November 2011).

Zhang, C.T. (2009) Advances in synthetic biology studies. *Bull. Nat. Sci. Found. China*, **2**, 65–69.

Further Reading

Legislation UK (2000) *The Genetically Modified Organisms(Contained Use) Regulations 2000* (ed. State TSo), HMSO, London.

Schmidt, M. (2009) *Do I Understand What I Can Create? Biosafety Issues in Synthetic Biology. The Technoscience and Its Societal Consequences*, Springer Academic Publishing.

Webmaster (2009) Science Times, http://www.antpedia.com/news/17/n-84517.htm (accessed 1 November 2011).

World Health Organization (2004) *Laboratory Biosafety Manual*, 3rd edn, Whole Health Organization, Geneva.

United Nations (2010) The Nagoya–Kuala Lumpur Protocol on Liability and Redress for Damage Resulting from Living Modified Organisms Born in Nagoya, Press Release, 12 October, United Nations, Nagoya.

GMO Compass (2010) GMO Compass 2010, http://www.un.org/News/briefings/docs/2011/110307_Biosafety.doc.htm (accessed 1 November 2011).

The Biological and Toxin Weapons Convention (2005) http://europa.eu/legislation_summaries/foreign_and_security_policy/cfsp_and_esdp_implementation/l33270_en.htm (accessed 1 November 2011).

League of Nations (1925) http://www.un.org/disarmament/WMD/Bio/1925GenevaProtocol.shtml (accessed 1 November 2011).

Annex A List of Biofuel Companies

Company name	Website	Country
AB Enzymes	www.abenzymes.com	Germany
Agrivida	www.agrivida.com/	USA
Algenol Biofuels	www.algenolbiofuels.com	USA
Altrabiofuels	www.altrabiofuels.com	USA (Ohio)
Amyris[a)]	www.amyrisbiotech.com	USA (Calif.)
Aquaflow Group	www.aquaflowgroup.com	New Zealand
BBI International	www.bbiinternational.com	USA (Colo.)
BEST Energies	www.bestenergies.com	Australia
BIOeCON	www.bio-e-con.com	Netherlands
Biofuel Advance Research and Development, LLC (BARD)	www.bardllc.com	USA (Pa.)
Biofuelbox	www.biofuelbox.com	USA (Calif.)
Biomaxx Systems	www.biomaxxsystems.com	Canada
Blue Marvel Energy	www.bluemarbleenergy.net	USA
Catalin	www.catilin.com	USA (Iowa)
Chevron Corp.	www.chevron.com	USA (Calif.)
Choren	www.choren.com	Germany
Codexis	www.codexis.com/wt/page/ bioindustrials	USA (Calif.)
Codon Devices	www.codondevices.com/	USA (Mass.)
EdeniQ	www.edeniq.com	USA (Calif.)
EIE Complex S.L.	www.eiec.eu	Spain
Force Technology	www.forcetechnology.com	Denmark

(Continued)

Synthetic Biology: Industrial and Environmental Applications, First Edition. Edited by Markus Schmidt.

Company name	Website	Country
Genifuel	www.genifuel.com	USA (Utah)
Genomatica	www.genomatica.com	USA (Calif.)
GEVO	www.gevo.com	USA (Colo.)
Global Bioenergies	www.global-bioenergies.com	France
Global Green Solutions	www.globalgreensolutionsinc.com	Canada
Green Hunter Energy Inc.	www.greenhunterenergy.com	USA (Tex.)
Green Star Products Inc.	www.greenstarusa.com	USA
International Energy, Inc.	www.internationalenergyinc.com	USA (D.C.)
Inventure Chemical Technology	www.inventurechem.com	USA (Wash.)
Iogen	www.iogen.ca	Canada
Joint Bioenergy Institute	www.jbei.org	USA (Calif.)
Kaiima Biofuels (Biofuels Int. Inc.)	www.kaiima.com	Israel
Kior	www.kior.com	USA/ Netherlands
LS9	www.ls9.com/	USA (Calif.)
Mascoma	www.mascoma.com	USA (Mass.)
Microbiogen Pty Ltd	www.microbiogen.com	Australia
National Biodiesel Pty Ltd	http://www.natbiogroup.com/ default.asp?id=19	Australia
Novozymes	www.novozymes.com	Denmark
OPX	www.opxbiotechnologies.com	USA (Colo.)
Rainbow Nation Renewable Fuels (National Biofuels Group South Africa)	http://www.natbiogroup.com	South Africa
Sapphire Energy	www.sapphireenergy.com	USA
Seambiotic	www.seambiotic.com	Israel
Sequesco	www.sequesco.com	USA (Calif.)
Solazyme	www.solazyme.com	USA (Calif.)
Solix Biofuels	www.solixbiofuels.com	USA (Colo.)
St1 Biofuels	www.st1.eu	Finland
Sunx Energy	www.sunxenergy.com	Canada

a) For lack of positive results, Amyris pulled out of the biofuel market in January 2012.

Annex B List of Bioremediation Companies

Company name	Website	Country
AB Enzymes	www.abenzymes.com	Germany
Adventus	www.adventusgroup.com	USA
Agroforestal San Remo	www.agroforestalsanremo.com	Venezuela
Alabaster Corp.	www.alabastercorp.com	USA
ALIBIO (Alianza con la Biosfera)	www.alibio.com.mx	Mexico
Arcadia Environmental	www.acadiaenvironmental.com	USA
Bauer	www.bauerumweltgruppe.com/	Germany
BioNaturTech	www.bionutratech.com	USA
Biopract	www.biopract.de	Germany
Bioremediate.com LLC	www.bioremediate.com	USA/Canada
Bioremediation Inc.	www.bioremediationinc.com	USA
Brockportmicrobiology	www.brockportmicrobiology.com	USA
Carus Corp.	www.caruscorporation.com	USA
Catalina Biosolutions	www.biocritters.com	USA
Clean Earth Inc.	www.cleanearthinc.com	USA
Ecotree	www.ecolotree.com	USA
Ecuavital-Biox	www.biorremediacion.org	Ecuador
EGEO Services Inc.	www.egeoservices.com	USA
EIEC Complex	www.eiec.eu	Spain
EM America	www.emamerica.com	USA
Enretech	www.enretech.co.za	South Africa

(*Continued*)

Company name	Website	Country
EnviroMondeSpilltech	www.enviromondespilltech.com	
EOS Remediation	www.eosremediation.com	USA (Canada/EU)
Flinders Bioremediation	www.flindersbioremediation.com.au	Australia
Genencor	www.genencor.com	USA
Genome Canada	www.genomecanada.ca	Canada
Golden Enviromental Services Inc.	www.goldenenviro.ca	USA (Calif.)
INAMBIO (Ingenieria Industria y Biorremediacion)	www.inambio.com	Mexico
International Enzymes	www.itstoyou.com/ internationalenzymes/index.htm	USA
Ivey International	www.iveyinternational.com	USA
MicroBac	www.micro-bac.com	USA
Microsorb	www.microsorb.com	USA
Natural Environmental Systems	www.naturalenviro.com	USA
OTI (Oil Treatment International)	www.za.oti.ag	South Africa
Pelorus	www.pelorusenbiotech.com	USA
Regenesis	www.regenesis.com	USA
Renovogen	www.renovogen.com	USA
Respirtek	www.respirtek.com	USA
RNAS	www.rnasinc.com	USA
Soil and Water Remediation	www.soilandwaterremediation.co.uk	UK

Index

Page numbers in *italics* refer to entries in tables or figures.

Synthetic Biology: Industrial and Environmental Applications, First Edition. Edited by Markus Schmidt.

c